Sewell Newhouse, Community Oneida

The Trapper's Guide

a manual of instructions for capturing all kinds of fur-bearing animals, and curing their skins - with observations on the fur-trade, hints on life in the woods and narratives of trapping and hunting excursions

Sewell Newhouse, Community Oneida

The Trapper's Guide

a manual of instructions for capturing all kinds of fur-bearing animals, and curing their skins - with observations on the fur-trade, hints on life in the woods and narratives of trapping and hunting excursions

ISBN/EAN: 9783744731645

Printed in Europe, USA, Canada, Australia, Japan

Cover: Foto ©berggeist007 / pixelio.de

More available books at **www.hansebooks.com**

THE TRAPPER'S GUIDE;

A

MANUAL OF INSTRUCTIONS

FOR

CAPTURING ALL KINDS OF FUR-BEARING ANIMALS, AND
CURING THEIR SKINS; WITH OBSERVATIONS ON THE
FUR-TRADE, HINTS ON LIFE IN THE WOODS,
AND NARRATIVES OF TRAPPING AND
HUNTING EXCURSIONS.

BY S. NEWHOUSE,

AND OTHER TRAPPERS AND SPORTSMEN.

THIRD EDITION.

EDITED BY THE ONEIDA COMMUNITY.

NEW YORK:
MASON, BAKER & PRATT,
142 & 144 GRAND ST.

1872.

Entered according to Act of Congress, in the year 1865, by
J. H. NOYES,
In the Clerk's Office of the District Court of the United States for the Southern District of New York.

CONTENTS.

INTRODUCTION.
PAGE 7—12.

Connection of Trapping with other Trades, 8. Observations on the Fur Trade, 9. Season for Trapping. 10. Statistics of the Fur Trade, 11.

THE TRAPPER'S ART.
PAGE 13—94.

I. PRELIMINARIES, 13—20. The Dead-fall, 13. Poisoning, 14. Shooting, 14. Steel Traps, 15. Requisites of a good Trap, 15. The Spring-pole, 17. The Sliding-pole, 18. The Clog, 18. Rule for Baiting, 19. Proper Outfit of Traps, 20. Profits of Trapping, 20.

II. CAPTURE OF ANIMALS, 21—78. The Muskrat, 21. The Mink, 23. The Marten, 25. The Sable, 26. The Ermine, 28. The Fisher, 30. The Fox, 32. The Otter, 35. The Sea Otter, 41. The Beaver, 42. The Wolf, 47. The Bear, 48. The Raccoon, 50. The Badger, 51. The Wild Cat or Bay Lynx, 53. The Lynx, 56. The Cougar, 58. The Jaguar, 59. The Lion, 61. The Tiger, 62. The Wolverine, 66. The Opossum, 67. The Skunk, 67. The Coypu Rat, 70. The Chinchilla, 71. The Squirrel, 71. The Woodchuck or Marmot, 72. The Gopher, 72. The Rat, 74. The Deer, 74. The Mouse, 77.

III. CURING SKINS, 79—83. General Rules, 79. Stretching Skins, 80. Board Stretcher, 80. Muskrat Stretcher, 81. Bow Stretcher, 82. Hoop Stretcher, 82.

IV. LIFE IN THE WOODS, 84—94. Outfit for a Campaign on Foot, 84. Outfit for an Excursion by Wagon or Boat, 86. Tent, 87. Stove and Furniture, 87. Bed and Bedding, 88. Camp Chest, 89. Cooking, 89. Jerked Meat, 91. Preparations against Insects, 91. Shanty, 92. Trapping Lines, 92. Conclusion, 93.

THE TRAPPER'S FOOD.
PAGE 95—107.

The Deer, 95. The Buffalo, 96. The Rocky Mountain Sheep or Bighorn, 98. The Argali, 99. The Prong-horn Antelope, 100. Squirrel Hunting, 101. The Ruffled Grouse, 101. Pinnated Grouse, 102. Sharptail Grouse, 103. Cock of the Plains, 103. Dusky Grouse, 104. Canada or Spruce Grouse, 104. White-tailed Ptarmigan, 105. Willow Ptarmigan, 105. European Grouse, 107. Water Fowl, 107.

FISHING IN AUTUMN AND WINTER.

PAGE 108—110.

Spearing Fish, 108. Fishing through the Ice, 109. Net-fishing in Winter, 109.

NOTES ON TRAPPING AND WOOD-CRAFT. By F. R., p. 111—121.
PLAN OF A TRAPPING CAMPAIGN. By PETER M. GUNTER, p. 122—125.
BOAT-BUILDING, p. 126—129.
SNOW-SHOES, p. 130.
OIL FOR FIRE-ARMS, p. 130.

NARRATIVES.

PAGE 131—205.

AN EVENING WITH AN OLD TRAPPER, 131—137. A YOUNG TRAPPER'S EXPERIENCE, 138—142. THE DEER HUNT, 143—145. MUSKRAT HUNTING, 146—158. AN AMATEUR IN THE NORTH WOODS, 159—174. TRAVELLING IN A CIRCLE, 175—180. AN EXPEDITION TO THE LAURENTIAN HILLS, p. 181—205.

APPENDIX.

PAGE 206—216.

HISTORY OF THE NEWHOUSE TRAP, 206—212. DESCRIPTION OF THE NEWHOUSE TRAP, 213—215. CONCLUSION, 216.

FULL PAGE ILLUSTRATIONS.

Portrait of S. Newhouse,		Frontispiece.
The Muskrat,	opposite page	21
The Mink,	" "	23
The Marten,	" "	25
The Sable,	" "	26
The Fisher,	" "	30
The Red Fox,	" "	32
The Otter,	" "	35
The Beaver,	" "	42
The Wolf,	" "	47
The Grizzly Bear,	" "	48
The Raccoon,	" "	50
The Badger,	" "	52
The Wild Cat,	" "	54
The Lynx,	" "	56
The Cougar,	" "	58
The Opossum,	" "	67
The Skunk,	" "	68
The Coypu Rat,	" "	70
The Chinchilla,	" "	71
The Woodchuck,	" "	72
The Rat,	" "	74
Family of Deer,	" "	76
Moose Yard,	" "	78
Mr. Newhouse's Tent and Stove,	" "	84
Log Shanty,	" "	92
Deer Breaking Cover,	" "	95
The Ruffed Grouse,	" "	101
The Wild Goose,	" "	107
Canoes,	" "	126
Snow-Shoes and mode of wearing them,	" "	130
Community Works, Willow Place, Oneida, N. Y.	" "	206

ILLUSTRATIONS IN THE TEXT.

The Otter Slide, page 37. The Claw Trap, 37. The Deer Trap, 76. Board Stretcher, 80. Muskrat Stretcher, 81. Shelter Tent, 85. Various sizes of the Newhouse Trap, 213-215.

INTRODUCTION.

BY THE EDITORS

THIS book was not originally designed for amateur sportsmen or for the reading public generally, but for practical workingmen who make or propose to make trapping a means of livelihood. The plan of it was suggested by a business necessity in the following manner:

Mr. S. Newhouse, a member of the Oneida Community, having become widely known as the maker of an excellent kind of steel-traps, and it being generally understood that the practical perfection of his traps is owing to the wisdom in wood-craft which he gained in early life by actual experience in trapping, he has often been applied to by his customers and others for information in regard to the best methods of capturing various animals. The most convenient way to answer such applications seemed to be to put his wisdom in print, and let it go abroad with his traps. In preparing for publication the material furnished by Mr. Newhouse for this purpose, the editors found new facts, inquiries, and written contributions relating to trapping and kindred pursuits crowding upon them, till the original idea of a small technical pamphlet swelled to the dimensions of the present work. The objects which they have finally aimed at have been, on the one hand, to furnish all the information needed in order to qualify a mere novice in trapping to enter upon the business intelligently and successfully; and on the other, to make an interesting book for all lovers of wood-craft, and for the reading public at large.

The character of the work, however, will be found to be

mainly in accordance with its original practical design; and it might properly be dedicated to poor men who are looking out for pleasant work and ways of making money; and especially to the pioneers of settlement and civilization in all parts of the world.

As honesty is always good policy, it is best also to confess here that the author and editors of this work are extensively engaged in the business of making steel-traps, and have an eye to their own interests, as well as to the interests of others, in this effort to help the business of trapping.

And here perhaps is the place to say something of the mutual relations of the several trades immediately concerned in the subject of this book, and of their importance in the machinery of universal business.

CONNECTION OF TRAPPING WITH OTHER TRADES.

Trapping, in the business series, is the intermediate link between trap making and the fur-trade. The trapper buys of the trap-maker and sells to the fur-dealer. The first furnishes him with weapons, and the second buys his spoils. Through the first, he is related to the manufacturers and merchants of iron and steel, who furnish materials for his traps, and to the hardware men who bring them to his door. Through the second, he connects with the fur-manufacturer, the hatter, and the clothes dealer, and sends supplies of comfort and luxury to the world of wealth and fashion.

Trapping and trap-making are directly subservient to the fur-trade. They may be said to be branches of it, or even to be its foundations. The fur-business expands as they prosper, and, *vice versa*, they prosper as the fur-business expands. The trapper and trap-maker watch the prices of furs, as the sailor watches the winds and the currents. When furs are high, trapping becomes active, and the trap-maker has his hands full of business. When furs are low, trapping declines, and the trap-maker has to dismiss his workmen.

The importance of the subservient trades, trapping and trap-making, can best be judged by looking at the statistics of the great fur-market for which they work. The fur-trade, everybody knows, is an immense business. The making of the weapons and the fighting may be out of sight, but the spoils of the war are seen by all. Many a colossal fortune, like John Jacob Astor's, has been founded on peltry; and many a frontier city, like St. Paul's, has been built up by the traffic that originates in the enterprise of the trapper and trap-maker.

OBSERVATIONS ON THE FUR-TRADE.

The following statements are made on the authority of members of large fur-dealing firms in the city of New York.

The yearly production of raw furs in the whole world is worth from seventeen to twenty millions of dollars, and the whole amount of the fur-trade, including manufactured goods, reaches a value of not less than one hundred millions.

The whole number of Muskrat skins alone, taken annually, is estimated at five or six millions; of which three millions are used in Germany.

Raw furs are divided by American dealers into two classes, viz., *shipping* furs, *i. e.* furs that are to be sent abroad; and *home* furs, or furs for use in this country. The leading articles among shipping furs are the Silver, Red and Cross Fox, Raccoon, Fisher, Wildcat and Skunk. Among home furs are the Mink, Opossum, House Cat, Wolf, and Marten. The Muskrat and other furs are classified under both heads.

Prices for shipping furs are regulated by the foreign demand.

The great fur-marts in Europe, are London, Leipsic, and Nijni Novgorod. At these points semi-annual sales (or fairs as they are termed), take place. The spring sales are most important. Here the representatives of the leading fur-houses from all parts of Europe meet to make their purchases during the months of March, April, and May.

The fur-trade of Leipsic is estimated at six and a half millions of dollars annually.

Raccoon fur is the great staple for Russia; Red Fox for Turkey and the oriental countries; Skunk for Poland and the adjacent provinces; Muskrat for Germany, France, and England.

New York is the great fur-mart in this country, and is the main depot of the shipping trade. There are no organized fur-companies at the present time. The business is carried on by private firms of large means and long experience. The New York Directory gives the names of more than one hundred furriers and fur-merchants, thirty of whom are wholesale dealers.

The leading fashionable fur for this country is the Mink; but the furs that are within the reach of the masses, and most worn, are the Muskrat and the Opossum.

The wearing of furs in this country is very little affected by *climate*, but is regulated almost entirely by *fashion*. In Europe, on the other hand, the state of the elements determines the extent of the call for furs as articles of clothing. Hence, notwithstanding the winters on both continents are growing milder, the demand for furs is continually increasing in this country, while in Europe it is falling off.

The more thickly settled parts of the United States show a large decrease in the "catch" of furs; but new territories are continually opening to the trapper; and though he moves from year to year farther north and west, the supply steadily keeps pace with the demand.

SEASON FOR TRAPPING.

All furs are best in winter; but trapping may be carried on to advantage for at least six months in the year, *i. e.* any time between the first of October and the middle of April. There is a period in the warm season, say from the first of May to the middle of September, when trapping is out of the question, as furs are worthless. The most trapping is done late in the fall and early in the spring.

The reason why furs become worthless in summer is, that all fur-bearing animals shed their coats, or at least lose the finest and thickest part of their fur as warm weather approaches; and have a new growth of it in the fall to protect them in winter. This whole process is indicated in the case of the Muskrat, and some other animals, by the color of the inside part of the skin. As summer approaches, it becomes

INTRODUCTION. 11

brown and dark. That is a sign that the best fur is gone. Afterwards it grows light-colored, and in winter when the fur is in the best condition it is altogether white. When the pelt is white it is called *prime* by the fur-dealers. The fur is then *glossy, thick,* and of the *richest color*, and the tails of such animals as the Mink, Marten, and Fisher are full and heavy. Beavers and Muskrats are not thoroughly prime till about the middle of winter. Other animals are prime about the first of November. There is probably some variation with the latitude, of the exact period at which furs become prime, the more northern being a little in advance. Trappers are liable to begin trapping too early in the season, consequently much poor fur is caught, which must be sold at low prices, and is unprofitable to the trapper, the fur-buyer and the manufacturer.

STATISTICS OF THE FUR-TRADE.

The following estimates of the annual production of all the fur countries in the world, were given in a volume on the fur-trade, published in 1864, by Heinrich Lomer, one of the principal fur-dealers of Leipzic. The total value is somewhat less than we have given on a previous page and is probably within the truth.

YEARLY PRODUCTION OF FURS THROUGHOUT THE WORLD.

Names of Furs.	Asia and Russian North America.	Middle Europe.	North America, and South America.	Russia, Sweden, Iceland, and Greenland.	Total.	Value in Thalers.
Sable	109,000		130,000	6,000	245,000	2,500,000
Mink			200,000	55,000	255,000	700,000
Pine Marten		120,000		60,000	180,000	840,000
Stone Marten		250,000		150,000	400,000	1,350,000
Fitch		380,000		220,000	600,000	600,000
Kolinsky or Tartar Marten	80,000				80,000	80,000
Ermine	350,000			50,000	400,000	100,000
Squirrels	6,000,000			1,000,000	7,000,000	1,400,000
Muskrat	150,000		2,850,000		3,000,000	1,500,000
German Marmot		250,000			250,000	2,500
Chinchillas			*100,000		100,000	80,000
Silver Fox	500		1,500		2,000	200,000
Cross Fox	5,000		4,800	100	10,000	77,000
Blue Fox				6,500	6,500	65,000
White Fox	54,000		8,000	23,000	85,000	85,000
Red Fox	45,000	140,000	60,000	85,000	330,000	700,000

YEARLY PRODUCTION OF FURS THROUGHOUT THE WORLD. — (Continued.)

Names of Furs.	Asia and Russian North America.	Middle Europe.	North America and South America.	Russia, Sweden, Iceland, and Greenland.	Total.	Value in Thalers.
Gray Fox			25,000		25,000	25,000
Kit Fox	30,000		10,000		40,000	40,000
Raccoon			600,000		600,000	600,000
Fisher or Pekan			12,500		12,500	100,000
Skunks			100,000		100,000	80,000
Opossum			250,000	*30,000	280,000	80,000
Marmot or Woodchuck	40,000	5,000	5,000	5,000	55,000	11,050
Bears	1,700		15,000	2,300	19,000	195,000
Lynx	15,000		26,000	9,000	50,000	175,000
Wolf	6,000	500	12,500	6,000	25,000	40,000
Buffalo			60,000		60,000	480,000
Wolverene	300		2,500	700	3,500	10,600
Badger		30,000	2,500	23,000	55,000	41,000
Beaver	30,000		130,000		160,000	575,000
Sea-Otter	1,200		300		1,500	200,000
Otter	4,000	12,000	20,000	9,000	45,000	305,000
Fur-Seals	25,000		*30,000		55,000	280,000
Seal	130,000	20,000	20,000 / 500,000	330,000	1,000,000	1,000,000
Coypu			*3,000,000		3,000,000	400,000
Hares	2,000,000	1,300,000		1,200,000	4,500,000	1,030,000
Rabbits		4,420,000	580,000		5,000,000	800,000
Cat	250,000	500,000	45,000	205,000	1,000,000	235,000
Lambskins	700,000	2,000,000		330,000	3,030,000	1,325,000
Monkey			*40,000		40,000	50,000
Lion and Tiger			*500		500	5,000
					32,050,500	17,456,650†

† Value in American coin, $12,721,152.50.

☞ In the above table the numbers marked with an * are the products of South America, Southern Asia, Africa, Australia, the islands adjacent to these countries, and the South Sea Islands.

THE TRAPPER'S ART.

By S. NEWHOUSE.

I. PRELIMINARIES.

WILD animals are taken for various reasons besides the value of their furs. Some are sought as articles of food; others are destroyed as nuisances. In these cases the methods of capture are not essential. Animals that are valuable for food may be run down by dogs, or shot by the rifle or fowling-piece; and nuisances may be destroyed by poison. But for the capture of fur-bearing animals, there is but one profitable method, namely, by *steel-traps*. Other methods were much used by trappers in old times, before good steel-traps were made; and are still used in semi-barbarous countries, where steel-traps are unknown, or cannot be had. I will briefly mention two or three of these methods, and the objections to them, and after that give my views of the true method.

THE DEAD-FALL.

This is a clumsy contrivance for killing animals, which can be made anywhere, with an axe and hard work. It consists of two large poles (or logs when set for bears and other large animals), placed one over the other and kept in place by four stakes, two on each side. The upper pole is raised at one end high enough above the lower to admit the entrance of the animal, and is kept up in that position by the familiar contrivance of the stick and spindle, or "figure four." A tight pen is made with sticks, brush, &c., on one side of this structure, at right angles to it, and the spindle projects obliquely into this pen, so that the bait attached to it is about eight inches beyond the side of the poles. The animal, to reach the bait, has to place his body between the poles and at right angles to them, and on pulling the spindle, springs the "figure four," and is crushed.

The objections to this contrivance are, first, that it takes a long time to make and set one, thus wasting the trapper's time; and second, that animals caught in this way lie exposed to the voracity of other animals, and are often torn in pieces before the trapper reaches them, which is not the case when animals are caught in steel-traps, properly set, as will be shown hereafter. Moreover, the dead-fall is very uncertain in its operation, and woodsmen who have become accustomed to good steel-traps, call it a "miserable toggle," not worth baiting when they find one ready made in the woods.

POISONING.

Animals are sometimes poisoned with strychnine. I have myself taken foxes in this way. I used about as much strychnine as would be contained in a percussion-cap, inclosed firmly in a piece of tallow as large as a chestnut, and left on the fox's bed. After swallowing such a dose, they rarely go more than three or four rods before they drop dead.

The objection to this method is, that it spoils the skin. Furriers say that the poison spreads through the whole body of the animal, and kills the life of the fur, so that they cannot work it profitably. Poison is used very little by woodsmen at the present time.

SHOOTING.

This method of killing fur-bearing animals, is still quite prevalent in some countries. It is said to be the principal method in Russia, and is not altogether disused in this country. But it is a very wasteful method. Fur-dealers and manufacturers consider skins that have been shot, especially by the fowling-piece, as hardly worth working. The holes that are made in the skin, whether by shot or bullets, are but a small part of the damage done to it. The shot that enter the body of the animal directly, are almost harmless compared with those that strike it obliquely, or graze across it. Every one of these grazing shot, however small, cuts a furrow in the fur, sometimes several inches in length, shaving every hair in its course as with a razor. Slits in the skin have to be cut out to the full extent of these furrows, and closed up or new pieces fitted in. Hence when the hunter brings his stock of skins to the experienced furrier, he is generally saluted with the question, " Are your furs shot, or trapped?." and if he has to answer, " They were shot," he finds the dealer quite indifferent about buying them at any price. The introduction of good steel-traps into Russia would probably add millions of dollars annually to the value of the furs taken in that vast territory.

STEEL-TRAPS.

The experience of modern trappers, after trying all other methods, and all kinds of new-fashioned traps, has led them almost unanimously to the conclusion, that the old steel-trap, when scientifically and faithfully made, is the surest and most economical means of capturing fur-bearing animals. Some of the reasons for this conclusion are these: Steel-traps can be easily transported; can be set in all situations on land or under water; can be easily concealed; can be tended in great numbers; can be combined by means of chain and ring with a variety of contrivances (hereafter to be described) for securing the animal caught from destruction by other animals, and from escape by self-amputation; and above all, the steel-trap *does no injury to the fur.*

And here I think it my duty as a true friend to the trapper, to give him the benefit of my experience and study in regard to the form and qualities of a good steel-trap, that he may be able to judge and choose the weapons of his warfare intelligently.

REQUISITES OF A GOOD TRAP.

The various sizes of traps adapted to different kinds of animals, of course require different forms and qualities, which will be spoken of in the proper places hereafter. But several of the essentials are the same in all good traps.

1. *The jaws should not be too thin and sharp-cornered.* Jaws made of sheet-iron, or of plates approaching to the thinness of sheet-iron, and having sharp edges, or, still worse, sharp teeth, will almost cut off an animal's leg by the bare force of the spring, if it is a strong one, and will always materially help an animal to gnaw or twist off his leg. And it should be known, that nearly all the animals that escape, get away by self-amputation.

2. *The pan should not be too large.* A large pan, filling nearly the whole space of the open jaws, may seem to increase the chances of an animal's being caught, by giving him more surface to tread upon in springing the trap. But there is a mistake in this. When an animal springs a trap by treading on the outer part of a large pan, his foot is near the jaw, and instead of being caught, is liable to be thrown out by the stroke of the jaw; whereas, when he treads on a small pan, his foot is nearly in the centre of the sweep of the jaws, and he is very sure to be seized far enough up on the leg to be well secured.

3. *The spring should be strong enough.* This is a matter for good judgment, that cannot well be explained here; but it is safe to say that very many traps, in consequence of false economy on the part of manufacturers, are furnished with springs that are too weak to secure strong and desperate animals.

4. *The spring should be tempered scientifically.* Many springs, in consequence of being badly tempered, "give down" in a little while, *i. e.*, lose their elasticity and close together; and others break in cold weather, or when set under water.

5. *The spring should be correctly proportioned and tapered.* Without this, the stronger it is and the better it is tempered, the more liable it is to break.

6. *The form of the jaws must be such as to give the bow of the spring a proper inclined plane to work upon.* In many traps, the angle at the shoulder of the jaws is so great, that even a strong spring will not hold a desperate animal.

7. *The adjustment of the spring and jaws must be such,*

that the jaws will lie flat when open. Otherwise the trap cannot well be secreted.

8. *The jaws must work easily in the posts.* For want of attention to this, many traps will not spring.

9. *The adjustment of all the parts and their actual working should be so inspected and tested that every trap shall be ready for use* — "*sure to go,*" *and sure to hold.* In consequence of the unfaithfulness of trap-makers in inspecting and testing their work, many a trapper, after lugging a weary back-load of traps into the wilderness, finds that a large portion of them have some "hitch" which either makes them worthless or requires a tug at tinkering before they can be made to do the poorest service.

German and English traps are almost universally liable to criticism on all the points above mentioned; and most of the traps made in this country fail in one or more of them.

In addition to the foregoing requisites, every trap should be furnished with a stout chain, faithfully welded, with ring and swivel. And let the trapper look well to the condition of the swivel. Many of the malleable iron swivels used by second-rate, careless manufacturers, will not turn at all; and many an animal escapes by twisting off chains that have these dead swivels.

In treating of the capture of particular animals, I shall have occasion to refer frequently to several contrivances that are used in connection with the fastening of steel-traps. I will therefore describe those contrivances here, once for all.

THE SPRING-POLE.

In taking several kinds of land animals, such as the marten and fisher, it is necessary to provide against their being devoured by other animals before the trapper reaches them, and also against their gnawing off their own legs, or breaking the chain of the trap by violence. The contrivance used for this purpose is called a *spring-pole*, and is prepared in the following manner: If a small tree can be found standing near the place where your trap is set, trim it and use it for a spring as it stands. If not, cut a pole of sufficient size and drive it firmly into the ground; bend down the top; fasten the chain-ring to it; and fasten the pole in its bent position by a notch or hook on a small tree or a stick driven into the ground. When the animal is caught, his struggles, pulling on the chain, unhook the pole, which flying up with a jerk, carries him into the air, out of the reach of prowlers, and in a condition that disables his attempts to escape by self-amputation or other violence. The size of the pole must be proportioned to the weight of the game which it is expected to lift.

THE SLIDING-POLE.

Animals of aquatic habits, when caught in traps, invariably plunge at once into deep water; and it is the object of the trapper, availing himself of this plunge, to drown his captive as soon as possible, in order to stop his violence, and keep him out of the reach of other animals. The weight of the trap and chain is usually sufficient for this purpose in the case of the muskrat. But in taking the larger amphibious animals, such as the beaver, the trapper uses a contrivance which is called the *sliding-pole*. It is prepared in the following manner: Cut a pole ten or twelve feet long, leaving branches enough on the small end to prevent the ring of the chain from slipping off. Place this pole near where you set your trap, in an inclined position, with its small end reaching into the deepest part of the stream, and its large end secured at the bank by a hook driven into the ground. Slip the ring of your chain on to this pole, and see that it is free to traverse down the whole length. When the animal is taken it plunges desperately into the region towards which the pole leads. The ring slides down to the end of the pole at the bottom of the stream, and, with a short chain, prevents the victim from rising to the surface or returning to the shore.

THE CLOG.

Some powerful and violent animals, if caught in a trap that is staked fast, will pull their legs off, or beat the trap in pieces; but if allowed to drag the trap about with a moderate weight attached, will behave more gently, or at least will not be able to get loose for want of purchase. The weight used in such cases is called a *clog*. It is usually a pole or stick of wood, of a size suited to the ring of the trap-chain, and to the size of the game. As the object of it is to encumber the animal, but not to hold it fast, the chain should be attached to it near one of its ends, so that it will not be likely to get fast among the rocks and bushes for a considerable time. The usual way is to slip the ring over the large end of the pole and fasten it with a wedge.

RULE FOR BAITING.

There is one general principle in regard to *baiting* animals that may as well be recorded and explained here, as it is applicable to all cases. It is this: *Never put bait on the pan of a trap*. The old-fashioned traps were always made with holes in the pan for strings to tie on bait; and many if not most novices in trapping imagine that the true way is to attract the animal's *nose* straight to the centre of action, by piling bait on the pan, as though it were expected to catch him by the head. The truth, however, is, that animals are very rarely taken by the head or the body, but almost always by a leg. When an animal pulls at a bait on the pan of a trap, he is not likely even to spring the trap, for he lifts in the wrong direction; and if he does spring it, the position of his head is such, especially if the bait is high on the pan, that he is pretty sure to give the jaws the slip. Besides, bait on the pan calls the attention of the wary animal to the trap; whereas he ought to be wholly diverted from it, and all signs of it obliterated. Bait should always be placed so that the animal in attempting to take it shall put a *foot* on the pan. This can be done in several ways, all of which will be explained in detail hereafter. But this general direction may be given for all cases that are not otherwise prescribed for: *Place the bait either on a stick above the trap, or in an inclosure so arranged that the animal will have to step over the trap to reach it.*

PROPER OUTFIT OF TRAPS.

In preparing for a trapping excursion, the novice naturally inquires how many traps he shall take along. If the question were simply how many traps he could *tend*, I should probably say from one to two hundred. But the main question really is, how many traps can he *carry?* If he is going on a marsh, lake, or river, where he can travel by boat, or into a region where he can carry his baggage by horse and wagon, he may take along all the traps he can tend, — the more the better. But if he is going by overland routes into the rough, woody regions where most game abounds, and consequently must carry his baggage on his back, he will probably find that seventy-five small traps, or an equivalent weight of large and small ones, will be as much as he will like to carry.

PROFITS OF TRAPPING.

The provident candidate for wood-craft will want to know what wages a man is likely to make at trapping. I will give him a few instances of what has been done, and then he may judge for himself. I have cleared seven dollars per day for a five weeks' trip. A man that once trapped with me, caught fifty-three muskrats in one night, which at present prices would be worth fifteen dollars and ninety cents. I know several men in Jefferson county (New York), who paid for good farms with furs that they caught within eight miles of home. It is not uncommon for two men to make five hundred dollars in a trapping season. But too much reliance must not be placed on these specimens. Good weather, good trapping-grounds, good traps, good judgment, and good luck must be combined, to secure good profits.

THE MUSKRAT.

II. CAPTURE OF ANIMALS.

It will be useful to the inexperienced trapper to have some account of the appearance and habits of each animal in connection with instructions for capturing it. Such information is often indispensable as the basis of plans and contrivances for capture. I shall confine myself to brief descriptions in common language, not attempting any thing scientific; and I shall avail myself of the help of books where my own observation and experience fail.

THE MUSKRAT OR MUSQUASH.

This is an animal of amphibious habits. Its head and body are from thirteen to fifteen inches in length. The tail is nine or ten inches long, two-edged, and for two thirds its length rudder-shaped, and covered with scales and thin, short hair, the edges being heavily fringed. The hind feet are slightly webbed; so that it can "feather the oar," as boatmen say, when they are brought forward in swimming. The color is brown above and ashy beneath. Muskrats are nocturnal in their habits; but are frequently seen swimming and feeding in the day time. They are excellent swimmers, and can go from ten to fifteen rods under water without breathing. Their natural food is grass and roots; but they will eat clams, mussels, flesh, corn, oats, wheat, apples, and many other vegetables. In open winters they sometimes find their way into farmers' cellars through drains, and make free with whatever they find in store. They thrive best in sluggish streams or ponds bordered with grass and flags. The roots of these plants are their chief support, and from the tops they construct their abodes. These structures are dome-shaped, and rise sometimes to the height of five or six feet. The entrances are at the bottom, under water; so that the inside of the houses are not exposed to the open air. The Muskrats live in them in winter, gathering into families of from six to ten members. Hundreds of these dwellings can be counted from a single point in many large marshes.

Muskrats have a curious method of travelling long distances under the ice. In their winter excursions to their feeding-grounds, which are frequently at great distances from their abodes, they take in breath at starting and remain under the water as long as they can. Then they rise up to the ice, and breathe out the air in their lungs, which remains in bubbles against the lower surface of the ice. They wait till this air recovers oxygen from the water and the ice, and then take it in again and go on till the operation has to be repeated. In this way they can travel almost any distance, and live any length of time under the ice.

The hunter sometimes takes advantage of this habit of the Muskrat, in the following manner: When the marshes and ponds where Muskrats abound are first frozen over and the ice is thin and clear, on striking into their houses with his hatchet for the purpose of setting his traps, he frequently sees a whole family plunge into the water and swim away under the ice. Following one of them for some distance, he sees him come up to renew his breath in the manner above described. After the animal has breathed against the ice, and before he has had time to take his bubble in again, the hunter strikes with his hatchet directly over him and drives him away from his breath. In this case he drowns in swimming a few rods, and the hunter, cutting a hole in the ice, takes him out. Mink, otter, and beaver travel under the ice in the same way; and hunters have frequently told me of taking otter in the manner I have described, when these animals visit the houses of the Muskrat for prey.

In summer, Muskrats live mostly in banks and in hollow trees that stand near a stream; and sometimes, for want of suitable marshes and ponds, they remain in the banks and trees through the winter. They are very prolific, bringing

THE MINK.

forth from six to nine at a birth, and three times a year. The first kittens also have one litter, which attain to about the size of house-rats in September. They have many enemies, such as the fox, wolf, lynx, otter, mink, and owl. They are found from the Atlantic to the Pacific, and from the Rio Grande to the Arctic Regions. But they do not inhabit the alluvial lands of Carolina, Georgia, Alabama, and Florida, though in other regions they live much further south.

The modes of capturing the Muskrat are various. One of them we have already seen. Another is by spearing, of which a fine example will be given in a subsequent article by Mr. Thacker. These methods are good at certain seasons and in certain conditions of the ice, &c.; but for general service there is no means of capture so reliable as the steel-trap. Traps should be set in the principal feeding places, playgrounds, and holes of the Muskrat, and generally about two inches under water. Bait is not necessary except when game is scarce and its signs not fresh. In that case you may bait with apples, parsnips, carrots, artichokes, white flag-roots, or even the flesh of the muskrat. The musk of this animal will sometimes draw effectually at long distances. The bait should be fastened to the end of a stick, and stuck over the trap about eight inches high, and in such a position that the animal will have to pass over the trap to take the bait. Care should be taken to fasten the trap to a stake in such a position that the chain will lead the captive into deep water and drown him. If he is allowed to entangle himself or by any means to get ashore, he will be very likely to gnaw or twist off a leg and get away.

THE MINK.

The Mink is found in the northern parts of America, Europe, and Asia. Its fur is very valuable, and in this country of late years has been the most popular kind. The Mink is carnivorous, and belongs to the *mustelidæ* or weasel family. It resembles the ferret and ermine. It is not amphibious like the muskrat, yet lives on the banks of streams and gets much of its food from them. It is of a dark brown color, has short legs, a long body and neck, and a bushy tail. In this country there are two varieties, which some naturalists have supposed were distinct species; one small, dark-colored, common in the Northern and Eastern States and Canada; the other larger, with lighter-colored, coarser, and less valuable fur, common in the Western and Southern States. The dark-colored variety measures from eleven to eighteen inches in length from the nose to the root of the tail, and has a tail from six to ten inches in length. The European and Asiatic Mink is a distinct species.

Mink are ramblers in their habits, except in the breeding season. They feed on fish, frogs, snakes, birds, mice, and muskrats; and the hen-roost frequently suffers from their depredations. They are very fond of speckled trout, and pretty sure to find out the streams where those fish abound. Their breeding season commences about the last of April, and the females bring forth from four to six at a litter. The young are hid by the mother till they attain nearly half their growth, as the males of this species, as well as of the marten, fisher, weasel, panther, and most carnivorous animals, destroy their young when they can find them.

Mink can be taken in steel-traps, either on land or in the water. Experts generally prefer to take them on land. The trap should be set near the bank of a stream. If one of their holes cannot be found, make a hole by the side of a root or a stump, or anywhere in the ground. Three sides of the cavity should be barricaded with stones, bark, or rotten wood, and the trap set at the entrance. The bait may be fish, birds, or the flesh of the muskrat, cut in small pieces; and it should be put into the cavity beyond the trap, so that the animal will have to step over the trap in taking the bait. The trap should be concealed by a covering of leaves, rotten vegetation, or, what is better, the feathers of some bird. In very cold weather the bait should be smoked to give it a stronger smell.

Mink can be attracted long distances by a scent that is prepared from the decomposition of eels, trout, or even minnows. These fishes are cut in small pieces, and put into a loosely-corked bottle, which is allowed to hang in the sunshine for two or three weeks in summer, when a sort of oil is formed

THE PINE MARTEN.

which emits a very strong odor. A few drops of this oil on the bait, or even on a stick without bait, will draw Mink very effectually.

The chain of the trap should be fastened to a spring-pole, strong enough to lift the animal, when caught, out of the reach of the fisher, fox, and other depredators; or if the trap is set near deep water, it may be attached to a sliding-pole, which will secure the game by drowning it. Both of these devices are fully described on pages 17 and 18.

THE MARTEN.

The Marten is found on this Continent from about north latitude forty degrees to the northern limits of the woods, or about sixty-eight degrees. On the Eastern Continent they inhabit all the North of Europe and Asia, except the treeless districts of the cold regions. The principal species are, the Pine Marten, which inhabits both continents, the Beech or Stone Marten of Europe, the Sable of Russia and Northern Asia, and the Japanese Sable. Naturalists class the fisher, also, with the Martens. The Russian Sable is the finest and most valuable of all the Martens. The Hudson's Bay and Lake Superior Martens are next in value. Those from Hudson's Bay, though really a variety of the American Pine Marten, are commonly called Hudson's Bay Sables, and their fur is known by that name in the markets of Europe.

The Marten belongs to the weasel family, and is carnivorous. It is about as large as the mink, and differs but little from the latter in form, save that its feet are larger and hairy to the toes, and its tail is somewhat larger and of a dark brown or black color. The fur of the American Pine Marten is generally of a yellowish brown, but varies greatly in color according to season, latitude, and locality. The Hudson's Bay and Lake Superior Martens are very dark-colored. The favorite haunts of these animals are the thick dark woods of the cold snowy regions. They are strictly arboreal in their *habitat*. They generally live in hollow trees, but occasionally they excavate dens in the ground. They feed on rabbits, birds, squirrels, mice, and other small animals; are fond of beech-nuts, and. it is said, resemble the bear in their fondness for honey. They are active climbers, and their small size enables them to pursue the gray squirrel and capture him in his hiding-places. They are, however, unable to cope in speed with the red squirrel or chickaree. They are not strictly nocturnal in their habits, as some have asserted, being frequently seen and killed in the daytime. Their breeding season begins in March or April, and they have from three to five young at a time, which are hidden from the males during infancy.

Sir John Richardson, the Arctic explorer, says that "particular races of Martens, distinguished by the fineness and dark color of their fur, appear to inhabit certain rocky districts."

Throughout the Hudson's Bay Territory there is a periodical disappearance of the Martens, which is very remarkable. It occurs, according to Bernard Rogan Ross, in decades, or thereabouts, with wonderful regularity, and it is not known what becomes of them. They are not found dead, and there is no evidence of their migration. The failure extends through the whole territory at the same time. In the seasons of their disappearance, the few that remain will scarcely touch bait. There seems to be a providential instinct in this by which the total destruction of the race is prevented.

Martens are taken in steel-traps by the same method as the mink. In winter, however, the traps should be set in hollow logs or trees, secured from the covering of snows, and concealed by the feathers of a bird. The Marten trappers of the Hudson's Bay Company commonly bait with fish-heads, pieces of flesh-meat, or, what they consider still better, the heads of wild fowl, which the natives gather for this purpose in autumn.

THE SABLE.

As I have already remarked, the Sable is closely allied to the martens. It is classed with them in Natural History, under the scientific name of *Martes Zibellina*. Two species are known: the *Martes Zibellina* or Russian Sable, and the Japanese Sable. The latter is marked with black on its legs and

The Russian Sable.

feet. It is thought by some of the Hudson's Bay Company's agents, that a marten exists in the northwestern part of British America, and in the late Russian Possessions, which, if not the same, is very closely allied to the Russian Sable. The Russian Sable is spread over a vast extent of territory, being found from the northern parts of European Russia eastward to Kamtschatka. Its size is about equal to that of the marten, being about eighteen inches in length exclusive of the tail. It is not very prolific, seldom bringing forth more than five at a birth, and generally only three. This takes place in March and April. They make their homes chiefly near the banks of rivers, and in the thickest parts of the woods. They usually live in holes which they burrow in the earth. These burrows are commonly made more secure by being dug among the roots of trees. Occasionally they make their nests in the hollows of trees, and there rear their young. Their nests are composed of moss, leaves, and dried grass, and are soft and warm. Their food varies with the season, and is partly animal and partly vegetable. In the summer, when hares and other small animals are wandering about, the Sable devours great numbers of them. But in winter, when these animals are confined in their retreats by the frost and snow, the Sable is said to feed on wild berries. It also hunts and devours the ermine and small weasels, and such birds as its agility enables it to seize. Sometimes, when other sources of food fail, it will follow the track of wolves and bears, and feed on the remnants of prey these animals may have left.

The fur of the Sable is in great request, and is the most beautiful and richly tinted of all the martens. The color is a rich brown, slightly mottled with white about the head, and having a gray tinge on the neck; it varies somewhat according to locality, and in some regions is very dark. The best skins are said to be obtained in Yakootsk, Kamtschatka, and Russian Lapland. Atkinson, in his "Travels in Asiatic Russia," says that Bagouzin, on Lake Baikal, is famed for its Sables. No skins have yet been found in any part of the world equal to them. The fur is of a deep jet black, with points of hair tipped with white. This constitutes their peculiar beauty.

From eighty to ninety dollars are sometimes demanded by the hunters for a single skin.

The Russian Sable is monopolized by the imperial family and nobility of that country. Only a few skins find their way into other countries. Some, however, are obtained privately in Siberia, by Jewish traders, and brought annually to the Leipzig fair. The fur of the Sable has the peculiarity of being fixed in the skin in such a manner that it will turn with equal freedom in all directions, and lies smoothly in whatever direction it may be pressed. The fur is rather long in proportion to the size of the animal, and extends down the limbs to the claws.

The best method of capturing the Sable is by the steel-trap, the same as I have already described for taking the mink and marten.

The Sable can be domesticated with success.

THE ERMINE.

Next in importance to the sable, amongst European furs, is that of the Ermine. The Ermine belongs to the weasel family, has the general weasel shape and appearance, and inhabits the northern parts of Europe and Asia. It is a small animal, measuring only about fourteen inches in total length, of which the tail occupies four inches. There is, however, considerable variation in the size of individuals. The Ermine is carnivorous and a most determined hunter. It preys on hares, rabbits, and all kinds of small quadrupeds, birds, and reptiles. It is very fond of rabbits, of which, especially the young, it destroys great numbers. The pheasant and partridge also suffer greatly from its predacity. It pursues its game with great pertinacity and rarely suffers it to escape. It is also a great plunderer of birds' nests of all kinds. Its favorite mode of attacking its prey is by fastening on the neck and sucking the blood of its victim. Wood, in his "Illustrated Natural History," gives the following account of the manner in which the hare is hunted by the Stoat or Ermine:

"Although tolerably swift of foot, it is entirely unable to cope with the great speed of the hare, an animal which frequently falls a victim to the Stoat. Yet it is enabled, by its great delicacy of scent and the singular endurance of its frame, to run down any hare on whose track it may have set itself, in spite of the long legs and wonderful speed of its prey. When pursued by a Stoat, the hare does not seem to put forward its strength as it does when it is followed by dogs, but as soon as it discovers the nature of its pursuer, seems to lose all energy, and hops lazily along as if its faculties were benumbed by some powerful agency. This strange lassitude, in whatever manner it may be produced, is of great service to the Stoat, in enabling it to secure an animal which might in a very few minutes place itself beyond the reach of danger, by running in a straight line.

"In this curious phenomenon, there are one or two points worthy of notice.

"Although the Stoat is physically less powerful than the hare, it yet is endowed with, and is conscious of, a moral superiority, which will at length attain its aim. The hare, on the other hand, is sensible of its weakness, and its instincts of conservation are much weaker than the destructive instinct of its pursuer. It must be conscious of its inferiority, or it would not run, but boldly face its enemy; for the hare is a fierce and determined fighter when it is matched against animals that are possessed of twenty times the muscular powers of the Stoat. But as soon as it has caught a glimpse of the fiery eyes of its persecutor, its faculties fail, and its senses become oppressed with that strange lethargy which is felt by many creatures when they meet the fixed gaze of the serpent's eye. A gentleman who once met with a dangerous adventure with a cobra, told me that the creature moved its head gently from side to side in front of his face, and that a strange and soothing influence began to creep over his senses, depriving him of the power of motion, but at the same time removing all sense of fear. So the hare seems to be influenced by a similar feeling, and to be enticed as it were to its fate, the senses of fear and pain benumbed, and the mere animal faculties surviving to be destroyed by the single bite."

The mink, marten, fisher, and other members of the weasel family, are said to exercise an influence on their prey similar to that above described.

The color of the Ermine in summer is a light reddish brown on the upper parts of the body, and lighter tinted or nearly white underneath. In winter, in the high northern latitudes, its fur changes to a delicate cream-colored white, on all parts of the body except the tip of the tail, which retains its black color and forms a fine contrast to the rest of the body. It is only in the coldest portions of Norway, Sweden, Russia, and Siberia that the Ermine becomes sufficiently blanched in winter to become of any commercial value. Russian Asia furnishes the greater portion of those caught. In England the Ermine, when in its summer coat, is commonly called the Stoat, and, on account of its predaceous habits, is thoroughly detested.

Ermine fur was formerly monopolized by the royal families and nobility of Europe, but now finds its way into the general markets.

The same general methods should be pursued in trapping the Ermine as in the case of the mink and marten.

THE FISHER.

This animal is usually called Pennant's Marten by the naturalists. From some hunters it also receives the name of Pekan. But in the fur-trade it is generally known as the Fisher. It is strictly a North American animal, ranging from the Atlantic to the Pacific, and from the mountains of North Carolina and Tennessee to the Great Slave Lake, and perhaps still further north.

The Fisher belongs to the weasel family, and resembles both the marten and the wolverene in its habits and general appearance, though much larger than the former and less than the latter. Its general shape is like that of the marten, but its head is more pointed, its ears are more rounded, its neck, legs, and feet are stouter in proportion, and its claws much stronger. An average, full-sized Fisher will measure about two feet from the nose to the root of the tail. Its tail is about fifteen inches in length. Its feet are large, short, and stout, and thickly covered with fur and hair. The color of its fur is dark brown or black, and its tail is black and bushy.

Fishers are found chiefly in the cold, snowy regions of the north, and are generally nocturnal in their habits, though less so than the fox. They do not live so exclusively in the

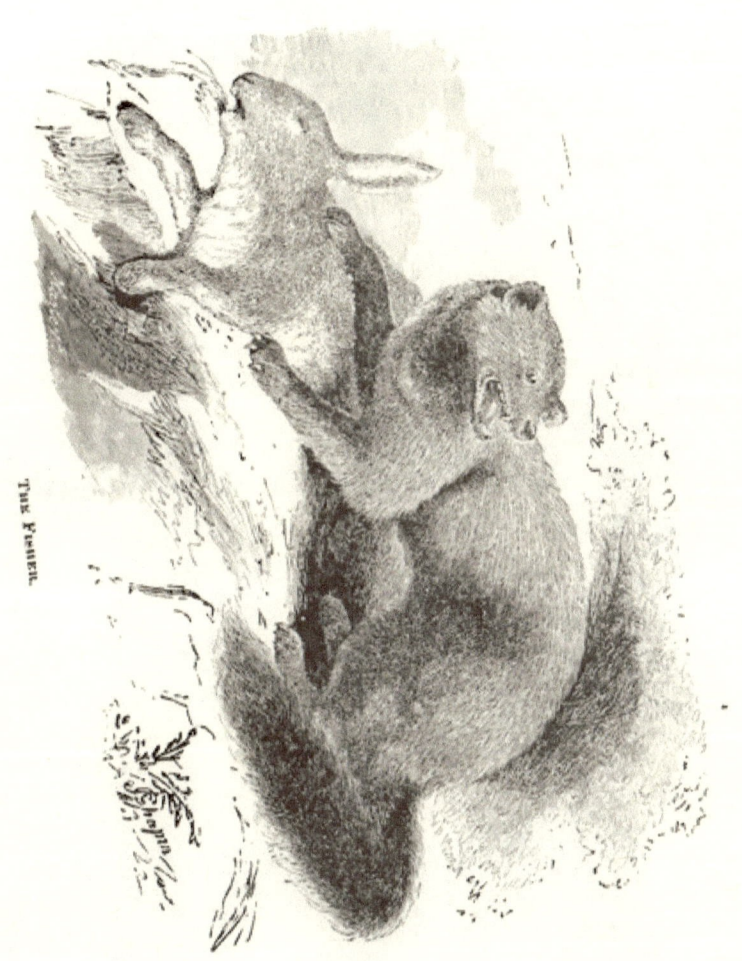
The Fisher.

woods as the marten, but their food is much the same. They prey on hares, raccoons, squirrels, grouse, mice, and small birds, and have been seen watching for fish, lying on a log that crossed the stream, with head inclined downward, ready for a plunge. They, however, prefer flesh-meat to fish. Their breeding season begins in March or April, and from two to four young are brought forth at a time. The young are hidden from the males in hollow trees at a considerable distance from the ground, until they are large enough to take care of themselves.

Fishers are taken in steel-traps by the same methods as the mink and marten. The barricade round the trap, however, should be stronger, and the entrance larger. The trap in all cases should be fastened to a spring-pole of sufficient strength to lift the animal clear from the ground, as it is pretty sure to gnaw off a leg or the pole, if left where it can touch the ground. The Hudson's Bay Company's trappers sometimes use the same methods in trapping the Fisher as those employed in fox trapping. Messrs. Holland and Gunter, trappers of many years' experience in the Laurentian Hills, of Canada West, describe their mode of trapping the Fisher as follows:—

"For capturing the Fisher, we always draw a trail composed of the oil of anise, assafœtida, and the musk of the muskrat, mixed with fish oil, and placed in a deerskin bag about the size of a mitten, pierced full of holes with a small awl. If drawn along the line of traps the scent is sure to attract the Fisher's attention, and when an animal once finds it, he will follow the trail till he comes to a trap. Mink are sometimes caught along trails of this kind; and it is a good plan to set a trap for wolves on the line, as they are likely to be attracted to and follow it. In setting the trap, we either place it in a hollow log, or build a strong house and place the trap at the entrance. In the latter case the bait should be placed in the back part of the house, about two feet from the door. The trap should be covered with finely powdered rotten wood. A spring-pole should be used, as all animals of the canine family will follow the trail and rob the traps. Deer-meat, muskrat-meat, or fish, make good bait for the fisher, marten, mink, or wolf."

The Fisher is an exceedingly powerful animal for its size, and will tear down wooden traps, or "dead-falls," with ease. It frequently annoys the trapper by robbing his marten-traps of their bait, or of the animals they have caught. Indeed, the marten-trappers of the Hudson's Bay Territory consider an old Fisher as great an infliction as a wolverene. It will follow a "line" of traps for miles, and visits them with exemplary regularity. The structure for taking the marten being too small to admit the entrance of a Fisher, he breaks in from behind, and thus secures the bait without getting into the trap.

THE FOX.

The members of the Fox or Vulpine genus are numerous. Foxes are distributed through all latitudes, but they are most abundant in the North. Naturalists recognize fourteen different species. On this continent we have the Red, the Cross, the Silver or Black, the Prairie, the Swift or Kit, the Gray, the Coast, and the Arctic species. Northern Asia is represented by the Black and Gray, the White, the Red, and the Kit; European Russia, Sweden, and Norway, by the Black and Gray, the Cross, the Blue, the White, and the Red; Middle Europe, by the Red; and Greenland by the Blue and the White. In Southern Africa the Asse or Caama, and in Northern Africa the Fennec or Zerda, belong to the Fox genus. Fur-dealers say that there are thirteen different varieties or species of the Fox in Russia.

The Fox is one of the most important of the fur-bearing animals. The most valuable, most beautiful, most rare, and most sought for of all the foxes, is the Silver Gray or Black. It is found in the high northern latitudes of both continents, but only about two thousand skins in all are annually obtained. The best ones bring at the London sales as much as two hundred dollars each. The Cross Fox is next in value. On this continent the Black, Cross, and Red Foxes vary greatly in color and marking, and in quality of fur. This is probably due to the hybridizing of the different species with each other. It is thought by some hunters that the Cross

The Red Fox.

Fox is a hybrid between the Red and the Black. It seems, however, to be a permanent variety.

The Fox belongs to the dog or wolf family, and is carnivorous in its habits. The different species closely resemble each other in size, form, habits, and mode of capture. They differ mainly in the color and quality of their fur, which varies, in consequence of difference in species and in climate, from the coarsest dog fur to the finest sable. The American Red Fox is the most common in this country, and in many parts of the United States is considered one of the worst robbers of the farmer's sheep-fold and hen-roost. The Red Fox of Europe, though closely resembling the American, is a different species.

Foxes feed on grouse, small birds, hares, rabbits, squirrels, muskrats, mice, fish, eggs; and some of them are remarkably fond of grapes, strawberries, and other ripe fruits. When pressed with hunger, they accept reptiles and carrion. Their modes of securing their prey are various. They generally seize their victim by creeping stealthily within springing distance, and pouncing on it like a cat; but they frequently pursue the rabbit and other game with the "long chase." Their senses of sight, smell, and hearing are very acute, and their speed is great. They are cunning, and their tricks to escape their enemies and secure their prey are very remarkable. The length of the Fox from the nose to the tip of the tail is about three feet, and its weight from fifteen to twenty-five pounds. The tail is large and bushy, and when wet retards their speed in running. Their breeding season is in February or March, and they bring forth from four to nine at a birth. They generally burrow and rear their young in the earth, but sometimes take up their abode in a hollow tree or log, or in a ledge of rocks.

Some of the most successful methods of catching the Fox are the following: —

To prevent the smell of iron from alarming the game, the trap should be thoroughly smeared with blood, which can be done by holding it under the neck of some bleeding animal and allowing it to dry. Or, for the same purpose, it may be heated and covered with beeswax, which at the proper temperature will readily run all over the trap and chain. It should be set near the haunts of the fox. A bed of ashes, chaff, or light earth should conceal the trap, and it should be fastened to a movable clog of six or eight pounds' weight, as directed on page 18. Wool, moss, leaves, or some other soft substance should be packed lightly under the pan and around the jaws. The surface of the earth in the neighborhood of the trap should be brushed with a quill or bush, so that all will seem natural. Scraps or small pieces of fried meat, rolled in honey, should be scattered over the bed of the trap, except where the pan is. Care should be taken to erase all footprints. To make the allurement doubly sure, obtain from the female of the dog, fox, or wolf the matrix in the season of coition, and preserve it in a quart of alcohol tightly corked. Leave a small portion of this preparation on something near the trap; and then, putting some of it on the bottom of your boots from time to time, strike large circles in two different directions, leading round to the trap. A piece of bloody meat may be drawn on these circles at the same time. The Fox, on striking this trail, will be very sure to follow it round to the trap and be caught.

Another method practised by woodmen is to set the trap in a spring that does not freeze over in winter, placing it about half an inch under water, and covering the space within the jaws with a piece of moss that rises above the water. A bait of meat should be placed in such a position that the Fox, in taking it, will be likely to put his foot on the moss, to prevent wetting it. The essence of the skunk is sometimes used in this case, in connection with the bait, with good effect; but most trappers prefer the preparation in alcohol, above mentioned.

Another good way is to obtain from the kennel of some tame Fox (if such can be found) a few quarts of loose earth taken from the place where the animal is accustomed to urinate. Set your trap in this material, and bait and smooth the bed as before. The Fox, cunning as he is, is not proof against such attractions.

THE OTTER.

THE OTTER.

The Otter is found in nearly all parts of the world. Eleven species, or at least varieties, have been noticed by naturalists. These inhabit the following countries: one species each in Europe, Island of Trinidad, Guiana, Brazil, Kamtschatka, Java, Madagascar, Pondicherry, Cape of Good Hope; and two species in North America. The principal species on this Continent, and the most important of all in the fur-trade, is the Canada or American Otter, scientific name *Lutra Canadensis*. The range of this Otter is from the Atlantic Ocean to the Pacific, and from the Gulf of Mexico to the shores of the Arctic Sea. The other North American species is the California Otter.

The Otter is aquatic in its habits, living in and near streams and getting its living from them. In appearance the Otter resembles a magnified mink. Its fur and color are much like those of the mink, and the lightening of the tints in age are the same in both. Its fur is short and thick. The under-fur is slightly waved and silky, and similar in texture to that of the beaver, but not so long. It has a silvery white shade. The color of the overlying hairs varies from a rich and glossy brownish black to a dark chestnut. The under parts are lighter than the upper. The Otter's ears are small and far apart; head broad and flat above; body thick and long; feet hard, short, and webbed; tail long, round, and toward the tip depressed, and flat beneath. The fur on the tail is the same as that on the body, but shorter. Its legs are apparently set upon the sides of its body, which gives it an awkward, waddling appearance when travelling on land. Otters frequently measure three feet and a half from the nose to the tip of the tail, and weigh from fifteen to twenty-five pounds.

They are excellent swimmers and divers, and can remain a long time under water. Their activity in this element enables them to take fish with the greatest ease. They even destroy fish in great numbers for the mere pleasure of killing them, when they do not require them for food. The speckled trout is their favorite game, and they frequent the clear rapid streams in search of this dainty. They are sometimes tamed and taught to drive fish into the net, and even to catch them and bring them ashore for their master. The Chinese or Indian Otter, called also the Nair-Nair, affords a good illustration of this capability. In every part of India the trained Otters are almost as common as the trained dogs in England.*

Otters burrow in the bank of streams, lining their nests with leaves and grass. The entrances to their abodes are under water. Their breeding season is in April or May, and the females bring forth from two to four young at a time.

They are gregarious and rambling in their habits, and have a singular practice of sliding down wet and muddy banks and icy slopes, apparently for sport. The places where they play in this manner are called "slides," and are found at intervals on all the streams and routes that they haunt. They

---

* The mode of instruction which is followed in the education of the Otter is simple, and is thus explained in Wood's *Illustrated Natural History:* "The creature is by degrees weaned from its usual fish diet, and taught to live almost wholly on bread and milk, the only fish-like article which it is permitted to see being a leathern caricature of the finny race, with which the young Otter is habituated to play as a kitten plays with a crumpled paper or a cork, which does temporary duty for a mouse. When the animal has accustomed itself to chase and catch the artificial fish, and to give it into the hand of its master, the teacher extends his instructions by drawing the leathern image smartly into the water by means of a string, and encouraging his pupil to plunge into the stream after the lure and bring it ashore. As soon as the young Otter yields the leathern prey, it is rewarded by some dainty morsel which its teacher is careful to keep at hand, and learns to connect the two circumstances together. Having become proficient in the preliminary instructions, the pupil is further tested by the substitution of a veritable, but a dead fish, in lieu of the manufactured article, and is taught to chase, capture, and yield the fish at the command of its master. A living fish is then affixed to a line in order to be brought by the Otter from the water in which it is permitted to swim; and lastly, the pupil is taught to pursue and capture living fish, which are thrown into the water before its eyes. The remaining point of instruction is to take the so-far trained animal to the water-side, and induce it to chase and bring to shore the inhabitants of the stream, as they swim unconstrained in their native element.

"When in pursuit of its finny prey, the Otter displays a grace and power which cannot be appreciated without ocular investigation. The animal glides through the water with such consummate ease and swiftness, and bends its pliant body with such flexible undulations, that the quick and wary fish are worsted in their own art, and fall easy victims to the Otter's superior aquatic powers. So easily does it glide into the water that no sound is heard, and scarcely a ripple is seen to mark the time or place of its entrance; and when it emerges upon the shore, it withdraws its body from the stream with the same noiseless ease that characterizes its entrance."

are frequently seen in troops of four or six wandering up or down a stream, and travelling for miles over hills and through swamps, from one stream or lake to the nearest point of another. In their rambles they make it a point to have a game of antics at every "slide" on their route. They are gone from home on excursions of this kind generally a week or ten days, and the trapper who knows their habits, is not disappointed if he does not catch them on their home-grounds the first or second night, but waits patiently for their return from their circuit.

Otter Slide.

I have shown on page 14 that the shooting of fur-bearing animals is a wasteful practice, because it injures the fur. It is especially wasteful in the case of aquatic animals, because they sink when shot in the water and generally are lost. Very few Otter are saved that are killed in this way.

Some trappers take the Otter with what is called a "claw-trap"—an instrument that springs like a common steel-trap, but strikes and kills the animal with claws or hooks. This trap should be set on the steepest "slides,"

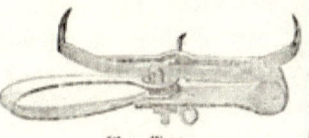

Claw Trap.

at about the middle of the descent and in the centre of the path, so that the Otter, in his game of sliding down hill shall spring the hooks and be struck in the centre of the body or breast. The trap must be carefully secreted.

But the common steel-trap is undoubtedly the best means of taking the Otter; and this instrument should be placed not on the middle of the "slide," but at the highest point of it, where the animal starts for his descent. The reason is, that at that point he is likely to be in a walking position, so as to be caught by his legs; whereas when he is on his way down the hill, he is sliding and rolling with his fore legs under his body, and is very apt to spring the trap with his breast or belly so as not to be seized by the jaws. Also the trap should be placed a little on one side of the central path of the "slide," because the legs of the Otter stand out on the sides of his body and are so far apart, that he is likely to put down his feet on each side of the trap and not in it, if it is set in the middle of the path. A small cavity should be made in the earth with a knife or hatchet, and the trap inserted so as to be nearly level with the path. Under the pan and around the jaws and springs there should be a light packing of leaves and moss. The top covering should be dry leaves of some evergreen or rotten wood broken very fine and brushed off smooth so as to appear natural. The trap should be fastened in the following manner: Cut a small tree of the size of the chain-ring and set it upright near enough to the path to assist in guiding the animal into the trap, supporting it in that position and securing the trap and game by withing or tying the top to another tree. The ring should be slipped on the butt and fastened by a wedge. After the trap is thus properly set, covered and fastened, a dry bush may be carelessly dropped in such a position as to turn the Otter in the right direction toward the trap. The whole apparatus should then be thoroughly drenched with water, which can be done by dipping an evergreen bough in the stream and sprinkling. Finally, perfume the place of the trap with a few drops of the fish-oil described on page 24. The musk of the Otter (which is an oil taken from two small glands, called oil-stones, lying next the skin on the belly of both sexes) may be added to complete the charm. The trapper, in his rounds of inspection, should be careful to keep at a proper distance from the trap while it is unsprung, so as not to leave any disturbing scent on the field of operations; for the Otter's sense of smell is incredibly delicate.

The art of taking Otter in the winter under the ice is not generally understood by trappers, and deserves an explanation. These animals do not hibernate, but travel about in winter as well as in summer. In the coldest weather they keep their feeding holes in the ice open, and are frequently seen near the edges, playing, sliding, and catching fish. They can be taken by the following process: Ascertain the depth of water at one of these holes, and cut a pole, suitable to the ring of the chain, and long enough to rise some distance above the ice when the butt is driven into the bottom. The ring of the chain should be slipped on the butt before it is driven, and should be free to traverse the length of the pole, except that a twig should be left near the lower end to prevent it from slipping off when you come to raise the trap. Two branches should emerge from one place toward the upper end, and should be left three or four inches long. Drive the pole so that these branches will be about eight inches below the ice, and fill into the fork of the branches with evergreens, so as to give the appearance of a bird's nest. Set your trap on this nest, and the Otter, climbing over it to assist him in emerging from the water, will spring it and be taken. Then he will make a desperate plunge to the bottom of the stream, and the ring of the chain sliding down on the pole, he will be unable to rise again and will drown. In this way many can be taken successively in a single trap. They travel mostly under the ice in winter, and in their rounds visit all the feeding holes on their way; and are often taken in traps set as above directed, when the holes are entirely frozen over.

During the winter the migrations of the Otter on land are toilsome, and it leaves a deep furrow or path in the snow. If a trap be set on this path the Otter is nearly certain to be caught, as it has a strong objection to opening new paths through the snow.

Other methods of trapping the Otter than those I have described are pursued by different trappers. Some trappers prefer to take them as they come out of the water near their "slides." It should be understood that Otters do not come ashore directly on to the "slide." They choose for their "slides" that part of the bank, of the stream or lake, which descends at a steep angle into deep water, so that when descending the "slide" they may plunge swiftly into the water without obstruction. In coming out of the water to go on the "slide," they choose a place where the water is shallow at the shore, and where they can walk up the bank easily. Hence, their place of exit is generally at a little distance from the "slide." The Canadian trappers, Holland and Gunter, describe their mode of trapping the Otter, as follows:—

"We set the trap close to the land, where the Otter comes out of the water to go on the 'slide.' We place the trap about three inches under water, and a little on one side of the path of the animal, so that the pan of the trap is about three inches from the centre of the path. The chain-ring of the trap we fasten to a pole fifteen feet long and one and a half inches in diameter. Then to the chain close to the pole we fasten a stone of about eight pounds' weight, to serve as an anchor; so that when the Otter is caught and makes for deep water, the stone sinks him to the bottom and he drowns. In cases where the water is too shallow to admit of setting the trap appropriately, an excavation should be made. If the water is too deep, place a flat stone or a piece of sunken wood under the trap. In all cases the trap should be set level; the anchor-stone and chain should be sunk under water; and the pole should be placed upright on one side of the path, in such a manner as to let the Otter get into deep water. We use the Newhouse Fox-Trap altogether for Otter."

Spencer J. Clark, of Oneida County, New York, who formerly trapped in Wisconsin, recommends setting the trap where the Otter comes out of the water, in the following position: The Otter swims to the shore, and as soon as his fore feet strike the ground his hind feet sink to the bottom, and he walks out erect. Find the point where the Otter's hind feet strike the bottom, and set the trap there. The advantages of this method are, first, the trap is in a position where it is not likely to be sprung, except by the Otter's feet; secondly, the trap can generally be set and visited in a boat without disturbing the shore, or leaving foot-prints and scent about the "slide." A sliding-pole should be used.

Other trappers prefer to set the trap several feet from the shore, on the path which the Otter takes in ascending to the top of the "slide." It should be set in the same manner as I have described for taking the animal when coming on to the "slide," on a preceding page. J. P. Hutchins recommends this method.

THE SEA-OTTER.

Along the northern shores of the Pacific Ocean, especially in Kamtschatka and Russian America, another species of Otter exists, called the Sea-Otter or Kalan. It is much larger than the fresh water Otter, weighing from sixty to eighty pounds. During the colder months of the year this Otter dwells by the sea-shores, where it is very active in the capture of marine fish. When warm weather approaches, the Kalan leaves the coasts, and with its mate proceeds up the rivers till it reaches the fresh water lakes of the interior, where it remains till cold weather again approaches. It is a rather scarce animal and not very prolific. The head and body measure from three to four feet in length. The tail is about seven inches long. Their food consists of fish, crustacea, mollusks, &c. This Otter haunts sea-washed rocks, around bays and estuaries, lives mostly in the water, and resembles the seals more than the Otters in its habits. It is very timid, and prefers the neighborhood of islands where it can find both food and shelter.

The fur of the Sea-Otter is very beautiful and of great value. Its color is variable, but the general hue is a rich black, slightly tinged with brown on the upper parts of the body, while the under portions and legs are of a lighter hue. About the head there is occasionally more or less white. The principal market for the skins is in China, where they are greatly prized by the official classes.

I cannot learn that any method of trapping the Sea-Otter

has ever been resorted to. They are usually hunted with boats, and shot. Audubon says that they are carefully approached by the boat, and when within a short distance are shot, and then harpooned by the bowsman before they sink. A careful study of their habits and haunts would probably indicate some method of taking them with the steel-trap, which would be more successful and inexpensive than any other.

THE BEAVER.

The Beaver belongs to the same family with the muskrat, and, like the latter, is amphibious. Indeed, these two species are so nearly alike, that a Beaver seems to be only a muskrat enormously enlarged. The body of the Beaver is thick, heavy, and squat; about two feet and a half long; weighing, when full grown, from sixty to eighty pounds. The tail is the most notable part of the animal. It measures from ten to twelve inches in length, and from three to four and a half inches in breadth. It is oval in shape, but flattened on the upper and under sides, and is covered with a species of hairy scales, which are set upon a thick, dusky skin. It is believed by trappers who have diligently watched the ways of this animal, that it uses its tail as a spade or trowel in working mud and sand. This member also answers the purpose of a prop, to help the animal stand erect while at work. It serves as both rudder and oar in swimming, being turned under the body at a right angle, and swung from side to side with great rapidity and power, the operation being like the sculling of a boat.

Beavers are not gregarious in summer, but become so at the approach of winter, when they build their huts and dams and gather their stores of food. Their huts are built first, generally in September, and are much like those of the muskrat, but larger and stronger. They rise out of the water, and have their entrances at the bottom. They are made to hold ten or twelve animals each. Some Beavers live on the banks of large rivers and lakes, and, having of course plenty of water, do not build dams, but have their holes in the banks,

The Beaver.

with their entrances under water, and their huts in front of them. These are called Bank Beavers, though they differ in nothing from their dam-building brethren. Those that live on small streams, where there is not water enough to surround their huts and protect their stores from freezing in winter, build dams to raise the water and create ponds suitable for their purpose. They commence by cutting down with their teeth trees of all sizes, from those of ten inches in diameter to the smallest brushwood. These are cut into pieces suitable for transportation by a single animal, and then are conveyed to the place chosen for a dam, the Beaver laying one paw over the timber, as he drags it along with his teeth. The smaller materials, such as mud, sticks, and stones, are carried between one of the fore-paws and the chin. The dams differ in shape according to the nature of the stream where they are built. In streams where the current is rapid or powerful, the dams are built with a convex curve up-stream, which strengthens them against the floods and the ordinary constant pressure of the stream. In streams where the water has but little motion, the dams are built straight across; and sometimes they have been observed with a curve down-stream. No special order or method is observed in building the dams, except that the work is carried on with a regular sweep, and all the parts are made of equal strength. They are frequently six or eight feet high, and from ten to thirty rods in length. The trees, resting on the bottom, are so mixed and filled in with mud, sticks, stones, leaves, and grass, that very little water escapes, except by running over the top; and the height is so uniform that the water drips evenly from one end to the other. After the dams are built, but before they are frozen over, the Beavers lay in their winter stores, which consist of the bark of the willow, aspen, poplar, birch, and alder. They fell these trees with their teeth, cut them up into short sections, and sink them in the water near their huts. In the winter, when their ponds are frozen over, they enter the water by the holes at the bottom of their huts, collect these sunken trees and take them to their dwellings, as they require them for food.

The breeding season of the Beaver commences in April or

May, and they have from two to four young ones at a birth. The young remain with their parents for three years. In the fourth year they start a new colony, and commence breeding, the parents assisting in building the new dam. This is probably the reason why so many dams are built one above another on the same stream. Several can frequently be seen from a single point, and they are generally so arranged that the water from one dam sets back to the next above.

The houses of the Beaver are built of the same materials as their dams. They are proportioned in size to the number of their inhabitants, which seldom exceed four old and six or eight young ones, though more than double that number have sometimes been found. Hearne, in his narrative of explorations in the Hudson's Bay country nearly a hundred years ago, relates an instance where the Indians of his party killed twelve old Beaver and twenty-five young and half-grown ones out of one house; and it was found, on examination, that several others had escaped. This house, however, was a very large one, and had near a dozen apartments under one roof, which, with two or three exceptions, had no communication with each other, except by water, and were probably occupied by separate families. In the spring, Beavers leave their houses, and roam about during the summer. On their return in the autumn, they repair their habitations for winter use. This is done by covering the outside with fresh mud. This operation is not finished until the frost has become pretty severe, as by this means the surface soon freezes as hard as stone, and prevents their great enemy, the wolverene, from disturbing them during the winter.

The food of the Beaver, beside the bark of the several kinds of trees I have mentioned, consists chiefly, in winter, of a large kind of root, somewhat resembling a cabbage-stalk, that grows at the bottom of lakes and rivers. In summer, they vary their diet by eating various kinds of herbage, and such berries as grow near their haunts.

Beavers are found in the northern parts of America, Europe, and Asia. They are generally supposed to belong to one species. They are most abundant on this Continent. Within a

CAPTURE OF ANIMALS. 45

recent period, Beavers were abundant in all the Northern, Middle, and Western States of the Union, as the large number of their dams, and the beautiful "beaver meadows" caused by the filling up of their ponds with alluvial matter, sufficiently indicate. But they retire at the approach of man; and the gradual clearing up and cultivation of the soil has driven them nearly all from the country. In the upper and lower provinces of Canada, however, they are still found in abundance.

There are several methods of taking Beaver in steel-traps. A few of the most successful I will endeavor to describe.

A full-grown family of Beavers, as I have said before, consists of the parents (male and female), their three-year-old offspring, the two-year-olds, and the yearlings, — four generations of four different sizes, occupying one hut, and doing business in one pond. When a trapper comes upon such a pond, or one that he has reason to believe is inhabited by a large number of Beavers, his object should be to take them all; and, in order to do this, he must conduct his operations so that when one Beaver is caught it will not have opportunity to alarm the rest; for otherwise the whole family may leave for parts unknown. His care should be directed therefore to two points, namely, first, to the setting of his traps in such a way as to take each Beaver while alone; and, secondly, to arrangements for drowning them as speedily as possible after they are taken. To secure the first point, he should not set his traps very near the dwelling of the Beavers, but should select places at some distance up the pond on some point or neck of land projecting into the stream, where the animals will pass and repass, but where each will be most likely to be alone. The trap should be set close to the shore, about three inches under water, and should be carefully secreted by a covering of some soft substance that will not interfere with its springing. For bait, a small portion of beaver-castor (a milky secretion found in glands near the testicles of the male Beaver) may be left on the bank near the trap. If the trapper's approach was made by land, all foot-prints should be erased by drenching with water. To secure the second point, the chain of the trap should be attached to a sliding-pole, in the manner described on page 18, which will lead the captured Beaver into deep water and drown him.

Beavers are sometimes taken by breaking away their dam, two inches below the surface, in one or two places, and setting traps in the breaches. They keep sentinels who examine their dams every night, and the least break is soon detected and put under repair; so that, with traps properly set, some of the Beavers will be likely to be taken while at work at this business. But, as the whole family is summoned out when a breach is considered dangerous, and as in any case several Beavers are likely to be engaged in a work of repair, the capture of one is almost sure to frighten away the rest, for which reason this method of capture should be generally discarded as impolitic.

The surest way of taking Beaver is by trapping in winter in the following manner: When their ponds are frozen over, make a hole in the ice about three feet across, near the shore and near a hut. Cut a tree of birch, poplar, or alder, about two inches in diameter; press the top together and shove the whole under the ice in such a direction that the Beavers will be likely to pass and repass it in going to and from their house. The butt of the tree should be fastened at the shore under the ice. Directly under the butt, about ten or twelve inches below, a platform should be prepared by driving stakes or by any other means that is convenient, on which the trap should be set. The chain ring should be attached as before to a *dry* sliding-pole. After the trap is set and secured, the hole in the ice should be filled up with snow and allowed to freeze. The Beaver, passing the newly cut tree and discovering its freshness, will proceed toward the butt for the purpose of securing the whole for food, and, in gnawing it off near the shore over the trap, will be likely to be taken. The reason why the sliding-pole should be dry is, that if it is green the remaining Beavers will be likely to gnaw it off and take it home with them, trap, Beaver and all, for the sake of the bark.

The Beaver is said to renew its breath, when travelling under the ice, in the same manner as the muskrat; and of

THE WOLF.

course might be caught at certain times in the way described on page 22.

THE WOLF.

There are many varieties of the Wolf, and they are found throughout North America, Europe, and Asia. They are substantially the same in form everywhere, but vary in color from black through shades of brown, fulvous, yellow, and gray, to white. The most common color is gray. They vary in size from the great White and Gray Wolves of the northern regions of America to the Coyote of the western plains. They inhabit chiefly unsettled and mountainous regions. They belong to the same family with the dog and fox. They are carnivorous, and combine both ferocity and cowardice in their character. Though lean and gaunt in appearance, they are fleet and powerful animals. They hunt mostly in packs, and destroy great numbers of deer in the stiff snows of winter, sometimes slaughtering whole herds in a single night. The sheepfold of the frontier farmer also suffers from their depredations. They feed on almost all the smaller animals they can overpower. Troops of them have been known to pursue and attack men. When hunting in packs and pressed with hunger they are bold and exceedingly ferocious. At other times, when roaming singly, they are sneaking and cowardly. The Gray Wolf of this country, which may be taken as the standard of size, is about four feet long from the point of the nose to the root of the tail; the length of tail being about seventeen inches. In the far north they are very large, sometimes measuring six and one half feet in total length, and weighing fifty pounds.

In North America the leading varieties are the Gray Wolf, the White Wolf, the Black Wolf, the Red Texan Wolf, and the Prairie Wolf or Coyote. In South America a Red Wolf is found in the marshy districts of the Rio de la Plata. In Europe there are Gray, Black, Brown, Red, and White Wolves. The latter are confined mostly to the Northern and Alpine regions. In Asia there are several varieties peculiar to that Continent.

It has been supposed by some that there is a variety on this

Continent which should properly be called the Giant Wolf. Old hunters say that occasionally there is seen in a pack of Wolves one that is larger and fleeter than its fellows. These are called "racers." They will run down a deer with ease. Whether such Wolves form a distinct variety, or are only overgrown individuals of the common varieties, has never been determined.

The breeding season of Wolves is in April or May, and they have from six to ten young at a time. They burrow in the ground or inhabit hollow logs or caves.

For capturing the Wolf by the steel-trap, the directions given in the first method of taking the fox should be followed, except that the honey should be left out, and the clog of the trap should be of fifteen or twenty pounds' weight. The small Prairie Wolf that is so troublesome to the western farmer can be captured in the same way. Care should always be taken to keep at a proper distance when looking after the trap, as the Wolf's sense of smell is very acute, and enables him to detect the foot-prints of the hunter with great sagacity.

The following plan for taking the Wolf is given by Peter M. Gunter, of Canada West: "Find two trees standing eighteen inches or two feet apart. Place the bait between the trees, and set a trap on each side of it. The traps should be smoked over hemlock or cedar boughs, to destroy any odor of iron. After being carefully set, the traps should be covered with finely powdered rotten wood. A clog of hard-wood of about twenty pounds' weight should be fastened to the chain of each trap. When all is arranged, rub some asafœtida on the trees to attract the attention of the wolves. If two trees cannot be found a suitable distance apart, lean two large logs against a tree where you wish to set your traps. It is better to use old logs, if lying about, than to make any fresh chopping."

THE BEAR.

The Bear family is very large. Its members inhabit nearly all parts of the globe, except Australia and the greater part, if not all, of Africa. They range through all latitudes from the equator to the poles. The following varieties and species

The Grizzly Bear.

have been described by naturalists: Polar Bear, Grizzly Bear, European Brown Bear, American Black Bear, Cinnamon Bear, Asiatic Bear, Siberian Bear, Spectacled Bear of South America, Thibetan Bear, Bornean Bear, and Malay Bear. The three latter are called Sun-Bears, from their habit of basking in the midday rays of the sun. They are the smallest members of the family, and live exclusively on vegetables.

Bears differ from each other, in consequence of differences of climate, more than almost any other animals. Those that inhabit the frozen wastes near the North Pole, or such high cold regions as the Rocky Mountains, are monsters of strength and ferocity; while those that inhabit warm countries are small, feeble, and inoffensive. The extremes of the scale are the Bornean Bear, which weighs less than one hundred pounds, and the great Polar Bear, which is thirteen feet in length, and weighs twenty-four hundred pounds. The American Black Bear is the species with which trappers have most to do. It is found in the western and northern parts of the United States and in the two provinces of Canada. Its weight when full grown is from three to six hundred pounds. The Cinnamon Bear of the Pacific coast is probably only a variety of this species.

Bears (except the Sun-Bears) are omnivorous, feeding indiscriminately on roots, berries, nuts, corn, oats, flesh, fish, and turtles. The farmer's calf-pasture, sheepfold, and hog-pen are frequently subject to their depredations. They are particularly fond of honey. They generally sleep through the coldest part of the winter. They bring forth their young in the months of May and June, and generally two at a time. The cubs are hid in caves or hollow trees till they are large enough to follow the dam, and then ramble about with her till the following spring.

The hunting of Bears with fire-arms, besides being objectionable on account of injury to the fur, is often dangerous business. They are very tenacious of life, and very bold and ferocious when wounded. A Grizzly Bear, shot by Captain Clark's party in the Rocky Mountain region, survived twenty minutes and swam half a mile after receiving ten balls in his body, four of which passed through his lungs and two through his heart! Records of Bear-hunting are full of perilous adventures, and those who engage in open battle with the great Grizzly Bear of the Rocky Mountains, rarely escape without loss of life or limb. But steel-traps of the right size, and properly managed, subdue these monsters with greater certainty than fire-arms, and without danger to the hunter.

In trapping for Bears, a place should be selected where three sides of an inclosure can be secured against the entrance of the animal, and one side left open. The experienced hunter usually chooses a spot where one log has fallen across another, making a pen in this shape ➤. The bait is placed at the inner angle, and the trap at the entrance in such a situation that the Bear has to pass over it to get at the bait. The trap should be covered with moss or leaves. Some think it best to put a small stick under the pan, strong enough to prevent the smaller animals, such as the raccoon and skunk, from springing the trap, but not so stiff as to support the heavy foot of the Bear. The chain of the trap should be fastened to a clog. (See page 18.) The weight of the clog for a Black Bear should be thirty pounds; for a Grizzly Bear, eighty pounds. The chain should not be more than eighteen inches in length, as the habit of the Bear, when caught, is to attempt to dash the trap in pieces against trees, logs, or rocks; and with a short chain, fastened to a heavy clog, he is unable to do this. The bait should be meat, and the Bear should be invited to the feast by the smell of honey or honey-comb, burnt on heated stones, near the trap. Bears seem to entertain no suspicion of a trap, and enter it as readily as a hog or an ox.

THE RACCOON.

The Raccoon is allied to the Bear family. It is found only on the Western Continent, where it is represented by two species: the Common Raccoon of the United States, and the Crab-eating Raccoon of the tropics. The former is spread over the greater part of North America from Texas to Hudson's Bay. On the Pacific coast it has been seen as far north

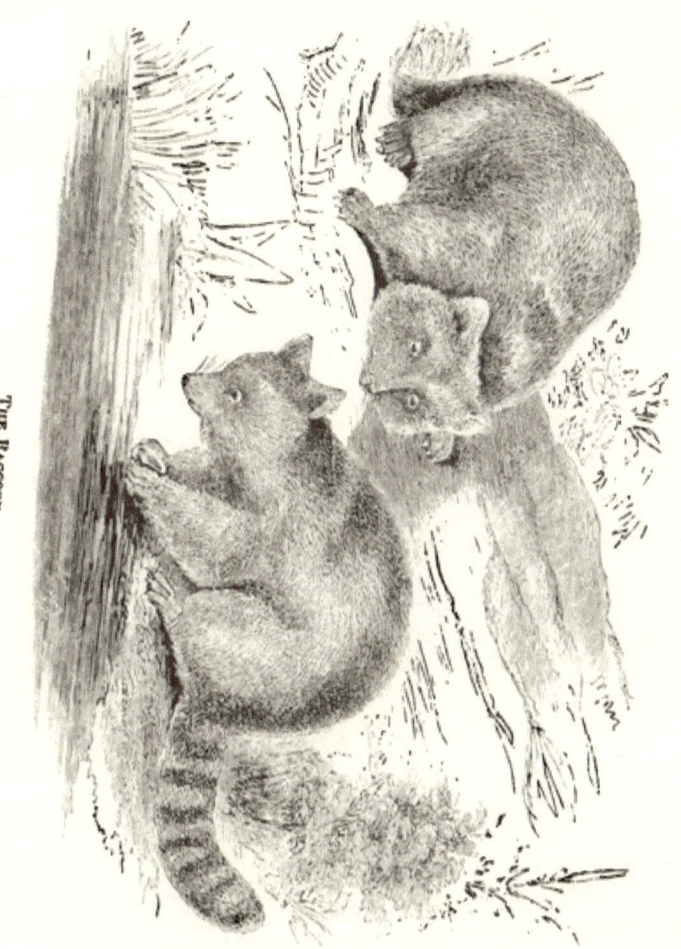

THE RACCOON.

as sixty degrees. The Crab-eating species is found from California and Texas to the 26th degree of south latitude.

The Common Raccoon is the one of principal interest to the trapper and fur-dealer. Its body is about two feet long, and is thick and stout like the badger's. Its head resembles that of the fox. Its tail is about a foot long, large, and bushy. The color of the whole is grayish white, streaked and barred with darker colors. In some of the Western States the Raccoon is of altogether a darker color, sometimes approaching to black. The Raccoon is nocturnal and omnivorous in its habits, and hibernates like the bear. It feeds on nuts, green corn, eggs, mice, frogs, turtles, fish, shell-fish, birds, &c., and frequently makes havoc in the poultry-yard. It is an excellent swimmer, and is fond of rambling about small streams and marshes in search of frogs, shell-fish, and turtles. It is also a good climber, and generally lives and rears its young in the hollow of a tree, with the entrance at a considerable height from the ground. Its breeding season is in April or May, and from four to six young are brought forth at a time.

Raccoons are sometimes taken by secreting traps in the paths which they make into corn-fields. Or traps may be set by the side of streams where they resort. In this case they should be baited with fresh fish; or, as some prefer, with salt cod-fish, roasted to give it a strong smell. They are not very cunning; and with their acute sense of smell, and their keen appetite for such provender, they rarely pass a trap thus baited without being taken.

THE BADGER.

This animal also belongs to the bear family. It is found in America, Europe, and Asia. Four species are recognized: the American Badger, the common Badger of Europe, the Indian Badger, and the Anakuma Badger of Japan. The European species is the most important in the fur-trade, furnishing 53,000 out of the 55,000 skins which annually find their way into the fur-markets.

Though spread over a large portion of the globe, the

Badger is nowhere numerous, except in a few localities on this Continent. It is omnivorous, feeding chiefly on roots, fruits, insects, and frogs. It also destroys the eggs and young of partridges, and other birds which build their nests on the ground. It is fond of the nests of wild bees, which it seeks out and robs with impunity, its tough hide being comparatively impervious to the stings of these insects. The Badger is a quiet, inoffensive animal, except when attacked, when it is a terrible antagonist to the dog or man who comes in contact with its sharp teeth and formidable jaws. Its length is about two feet six inches from the nose to the root of the tail. The tail is short. The head is small, flat, and has a long snout. The height at the shoulder is about eleven inches. The body is broad and flat, as though compressed. The legs are sturdy and powerful. The feet, before and behind, have each five toes strongly set in the flesh, and armed with powerful, compressed claws, adapted to burrowing in the ground, digging for roots, and unearthing the marmot, ground-squirrel, and other small, burrowing animals.

The Badger chooses the most solitary woods for its residence. It lives in burrows, where it makes its nest and rears its young. When pursued, it commences digging in the earth, and, if pressed too closely to be able to hide by burrowing, it makes a hole large enough to cover its body, backs into it, and faces its pursuers with claws drawn in an attitude of defiance; and woe to the dog that attempts to dislodge it from its fort! If it has time to get its body fairly buried, it is secure from any dog, or even a man with a shovel, as it digs so rapidly that it will work its way into the earth faster than dog or man can follow.

The fur of the Badger, when properly dressed, is said to make the best pistol furniture, and the coarser hairs are used for the fine brushes of the oil-painter. The hairs of the upper part of the Badger's body individually have three distinct colors: yellowish-white at the root, black in the middle, and ashy-gray at the end. This gives a uniform sandy-gray color to all the upper parts. The tail is furnished with long, coarse hair of the same color and quality. The throat, under parts,

The European Badger.

and legs are covered with shorter hair of a uniform deep-black.

The female Badger brings forth from three to five young in the early spring, suckles them for five or six weeks, and then turns them off to shift for themselves.

The American Badger differs considerably from the European species, to which the foregoing description applies. Its snout is less attenuated, though its head is equally long. The claws of its fore-feet are much longer in proportion, and its tail shorter. Its fur, both in color and quality, is different. It is also more carnivorous. Audubon describes its color and fur as follows: "Hair on the back, at the roots dark-gray, then light-yellow for two thirds its length, then black and broadly tipped with white, giving it in winter a hoary-gray appearance; but in summer it makes a near approach to yellowish-brown. The eyes are bright, and piercing black. . . . . There is a white stripe running from the nose over the forehead and along the middle of the neck to the shoulder. Legs, blackish-brown; chin and throat, dull-white; the remainder of the under surface, yellowish-white; tail, yellowish-brown." The fur on the back in winter is three inches long, dense and handsome. The body is broad, low, and flat.

The American Badger is abundant on the plains of the buffalo region of Dakotah and Nebraska, and in the timberless regions in the neighborhood of the Yakima River, Washington Territory. It is not found east of the Mississippi. It has been traced as far north as latitude fifty-eight degrees, and south into Mexico, where a distinct variety is found.

Badgers can be taken by setting traps at the mouths of their holes, or by the method prescribed on a preceding page for taking the raccoon. The trap should be carefully concealed, as the Badger is somewhat cunning, and disposed to be suspicious of such apparatus near his haunts.

THE WILD CAT OR BAY LYNX.

The American Wild Cat is a species of lynx. It is about thirty inches long, with a tail of five or six inches, and weighs from seventeen to twenty pounds. Its general color above and on the sides is a pale reddish brown, overlaid with grayish; the latter color most prevalent in spring and summer. The throat is surrounded with a ruff or collar of long hair. The under parts are light-colored and spotted. On the sides are a few obscure dark spots, and indistinct longitudinal lines along the middle of the back. The tail is marked with a small black patch above at the end, and with half rings on its upper surface. The inner surface of the ear is black, with white patch. The legs are long, the soles of the feet naked, and the hind-feet are partially webbed. The fur is moderately full and soft. The ears have a pencil of dark hairs in winter.

A variety of the American Wild Cat exists west of the Rocky Mountains, which was called by the early settlers in that region the Red Cat. Its color is somewhat darker than the common variety, being a rich chestnut-brown on the back; sides and throat, a little paler. Fur soft and full.

The Wild Cat is cowardly, rarely attacking any thing larger than a hare or young pig or lamb. The pioneer's henroost sometimes suffers from its nocturnal visitations. It feeds on grouse, partridges, squirrels, mice, and other small birds and quadrupeds. It is fond of the dark, thick cedar swamps, where it preys on rabbits, pouncing on them from an overhanging cliff or tree. In the Southern States, it frequents the swamps and canebrakes bordering on rivers and lakes, and also the briery thickets which grow up on old fields and deserted cotton lands. In dry seasons, or during the sultry weather of summer, it explores the courses of small streams, to feed on the fish that are left in the deep holes as the water dries up.

Wild Cats are taken in the same way as raccoons or minks, by baiting with meat, and covering the trap smoothly over. The best way is to find a place where they have killed a hare, grouse, or other game, and have left a part of the flesh for a second meal. Set your trap there, and you will be pretty sure of a visit.

The European Wild Cat is a distinct animal from the Bay Lynx. Goodrich, in his "Illustrated Natural History," gives

The American Wild Cat.

the following account of this Cat and of its relations to the common Cat: —

"There are many kinds of Wild Cat, but that from which the domestic Cat is supposed to have sprung is called the *Common European Wild Cat*, and is found in most parts of that quarter of the globe, as well as in Asia and Africa; it is also sometimes met with in this country. When America was first discovered, this species, either tame or wild, was not found here; all our domestic Cats, as well as the wild ones occasionally found in the woods, are the descendants of those brought hither by the Europeans. The Wild Cats of the European Continent are either the descendants of the original races that have continued untamed from the beginning, or of domesticated cats that have wandered from their homes, and, living apart from man, have relapsed into barbarism. It is said that the wild and tame Cats, in their wanderings, sometimes meet; when this is the case, the females of the tame breed are well treated by the savage Cats, but the males are rudely set upon and sometimes torn in pieces. The wild and tame Cats sometimes breed together, and produce the kind called *Tiger Cats*. Some authors hold that the Wild Cat is a distinct species, because its tail is shorter and more bushy than that of the domestic Cat; but this opinion seems not well founded, for still greater differences are found in dogs which are acknowledged to be of the same race."

The European Wild Cat is common in France, Germany, Russia, Hungary, and some other parts of Europe, and is found in Northern Asia and Nepaul. It was formerly found in England, and a few yet linger among the hills of Scotland. It resembles the tame Cat, but is rather larger and more robust, and has a more savage aspect. Its fur is long, soft, and thick. Its color is gray, darker on the back than below, with a blackish stripe along the back and paler curved stripes on the sides. It is a very shy animal; lurks in the woods and preys on hares, squirrels, and birds, and is for the most part nocturnal in its habits. It makes its home in clefts among rocks or in hollow trees. The female brings forth from three to six young at a time. A full-grown male is about two feet and a half long from the nose to the root of the tail; with a tail of considerable length. The female is smaller.

This Wild Cat is of great strength, and when pursued and hard pressed exhibits daring and ferocity in an extraordinary degree. When caught in a trap they fly without hesitation at any person who approaches them, without waiting to be assailed. The directions given for trapping the American Wild Cat are appropriate for the capture of this species. St. John, the author of a work on "Highland Sports," gives the following plan for taking them: "Like other vermin, the Wild Cat haunts the shores of the lakes and rivers, and it is, therefore, easy to know where to lay a trap for them. Having caught and killed one of the colony, the rest of them are sure to be taken, if the body of their slain relative is left in some place not far from their usual hunting-ground, and surrounded with traps, as every Wild Cat who passes within a considerable distance of the place will surely come to it."

THE LYNX.

There are several species of Lynx. The Canada Lynx and the European Lynx are the most important to the trapper and fur-dealer. The former inhabits North America from the latitude of Northern New York to the northern limits of the woods, or within the Arctic Circle. It is not found in the Mississippi Valley, but occurs west of the Rocky Mountains, and is supposed to exist in the northeastern part of Asia. Its size is between that of a fox and a wolf. Its length from the tip of the nose to the tip of the tail is about three feet. The tail is shorter than the head, and is densely furred and tipped with black. Its feet are large, thickly covered with fur, and armed with strong claws. The ears are pointed, not large, and tipped with a pencil of long black hairs. The color in winter is a silver-gray on the back, paling towards the belly, which is sometimes white. A rufous under-shade mixes with the tints. It has a ruff on the sides of the neck and under the throat. In winter its fur is long and silky. The average weight of this Lynx is about twenty-five pounds.

The Canada Lynx lives in the darkest woods and swamps, preying on hares, mice, squirrels, grouse, and smaller birds, and rarely attacking the deer. When pressed with hunger

THE CANADA LYNX.

it prowls about the pioneer's cabin in search of lambs, pigs, and poultry. It is an active climber, and frequently seizes its prey by pouncing upon it from an overhanging tree; at other times it crawls stealthily like a cat within springing distance, or leaps upon it from a cliff. It pursues birds to the tops of the loftiest trees, and kills fish in the streams. It also feeds on carrion, and, when pressed with hunger, on its own kind. It is said to have a strong passion for perfumes, particularly the castoreum of the beaver. This is the principal scent or "medicine" used by trappers in capturing the Lynx. The female brings forth generally two young ones at a time, and hides them in hollow trees or caves till they are large enough to follow her.

The Canada Lynx is a stupid animal and easily caught. It readily enters a trap that is properly set and baited with meat. The general directions already given for trapping various carnivorous animals are applicable in this case. The Hudson's Bay Company's trappers practice the following method, according to Bernard Rogan Ross: The trap is covered, inside the jaws, with a well-fitting "pallet" of birch bark. On the pallet a piece of hair skin, well rubbed with the "medicine" or scent, is tied. The trap is then placed indifferently either under or on the snow. The Lynx, scenting his favorite perfume, endeavors to withdraw the skin with his paw, and consequently springs the trap. It does not, like most of the fur-bearing animals, make violent efforts to escape, or drag the trap to a distance; it generally lies down until aroused by the approach of the hunter, when, instead of attempting to escape by flight, it springs at him.

The European Lynx closely resembles the Canada species; its habits are also similar. Its fur is valuable. Its general color is a dull reddish gray above, whitish below, mottled with black. On the sides are dark oblong patches. In winter the fur is longer and lighter-colored than in summer. The keenness of its sight has long been proverbial. It is found from the Pyrenees to the far North, and throughout Northern Asia. The directions given for trapping the Canada Lynx are sufficient in the case of this species.

THE COUGAR OR AMERICAN PANTHER.

This animal is one of the largest of the cat family that exists on the Western Continent, being rivaled only by the jaguar. It inhabits every latitude from Canada to Patagonia. In different localities it receives different names and varies somewhat in size. In the United States, east of the Rocky Mountains, it is commonly called the Panther, and sometimes the Catamount; on the west coast it is called the California Lion; in South America its common name is Puma. Cougar, however, is the scientific and proper name. The true Panther is confined to the Eastern Continent; and is a variety of the leopard, being found mostly in Asia. In the north, Cougars prefer for their retreat ledges of rock inaccessible to man, called by hunters *panther ledges*. They appear rarely by daylight, except when pressed for food, but conceal themselves behind rocks and fallen trees till evening. In South America their favorite haunts are the vast grassy plains, where they destroy great numbers of wild cattle.

Full grown Panthers killed in northern New York have been known to measure over eleven feet from the nose to the tip of the tail, being about twenty-eight inches high, and weighing nearly two hundred pounds. Their color is a reddish-brown above, shading into a lighter color underneath. They are armed with sharp teeth and long, heavy claws. They feed chiefly on deer, crawling stealthily to within springing distance, or watching on some cliff or tree, and pouncing like a cat on their prey. Their activity enables them to take the deer with ease. It is asserted by hunters that each Panther destroys as many as two deer per week, and a pair of Panthers have been known to attack and kill a full-grown moose. In newly settled countries, they frequently carry off young cattle and sheep. They are good climbers and readily take to a tree when pursued by dogs, from which they can easily be brought down by the rifle. This is the most common way of taking them. They are cowardly, and rarely attack a man unless wounded, when they are dangerous.

The Cougar.

The best way to take Panthers with steel-traps is to find where they have killed a deer or other animal, and left part of the carcass. Secrete the trap near the remains, and you will catch them when they return for a second meal. They seldom leave the vicinity of an animal they have killed, till it is all devoured. The same is true of all the large animals of the cat kind, such as the lion, tiger, leopard, jaguar, &c.

THE JAGUAR.

Like the cougar, this is an exclusively American animal. Though scarcely equalling the cougar in extreme length, the Jaguar is stouter and more formidable. It is found from Louisiana to Buenos Ayres. This animal has a large head, a robust body, and is very ferocious. Its usual size is about three fourths that of the tiger. Humboldt, however, states that he saw Jaguars which in length surpassed that of all the tigers of Asia which he had seen in the collections of Europe. The Jaguar is sometimes called the American tiger. Their favorite haunts are the swamps and jungles of tropical America. There they subsist on monkeys, capabyras or water-hogs, tapirs, peccaries, birds, turtles and turtle eggs, lizards, fish, shell-fish, and insects. Emerging from these haunts into the more open country, they prey upon deer, horses, cattle, sheep, and farm stock. In the early days of the settlement of South America the Jaguar was one of the greatest scourges the settlers had to meet. They haunted the clearings and plantations and devoured horses, cattle, and sheep without mercy. Nor were the settlers themselves and their children free from their attack. For many years where Jaguars abounded the settlers had an arduous warfare before they could exterminate the ferocious marauders, or drive them from the vicinity of their habitations.

The Jaguar is a cautious and suspicious animal. It never makes an open attack on man or beast. It approaches its prey stealthily, and pounces upon it from some hiding-place, or some position of advantage. It will follow a herd of animals for many miles in hopes of securing a straggler; and always chooses the hindmost animal, in order that if turned upon, it may escape with its prey the more easily. In this way it pursues men. A Jaguar has been known to follow the track of travellers for days together, only daring to show itself at rare intervals. A full grown Jaguar is an animal of enormous strength, and will kill and drag off a horse or ox without difficulty. They commit vast havoc among the horses which band together in great herds on the plains of South America. Full grown colts and calves are their favorite prey. Goodrich, in his Natural History, describes their operations as follows: "Frequently two Jaguars will combine to master the more powerful brutes. Some of them lie in wait around the salt-licks, and attack the animals that resort to these places. Their habit is to conceal themselves behind some bush, or on the trunk of a fallen tree: here they will lie, silent and motionless, for hours, patiently waiting for their victims. When they see a deer, or a mule, or mustang approaching, the eyes dilate, the hair rises along the back, the tail moves to and fro, and every limb quivers. When the unsuspecting prey comes within his reach, the monster bounds like a thunderbolt upon him. He fixes his teeth in his neck and his claws in the loins, and though the dismayed and aggravated victim flies, and rears, and essays to throw off his terrible rider, it is all in vain. His strength is soon exhausted, and he sinks to the earth an easy prey to his destroyer. The Jaguar, growling and roaring in triumph, already tears his flesh while yet the agonies of death are upon him. When his hunger is appeased he covers the remains of the carcass with leaves, sticks, and earth, to protect them from the vultures; and either remains watching near at hand or retires for a time till appetite revives, when he returns to complete his carnival." The Jaguar makes its attack upon the larger quadrupeds by springing upon their shoulders. Then placing one paw on the back of the head and another on the muzzle, with a single wrench it dislocates the neck. The smaller animals it lays dead with a stroke of its paw.

The Jaguar in external appearance and in habits closely resembles the leopard of the Old World. The female produces two at a birth. The ground color of a full-grown animal is yellow, marked with open figures of a rounded-angular form. In each of these figures are one or more black spots. The figures are arranged longitudinally and nearly parallel along the body. The belly is almost white. There is considerable variation in color among Jaguars, some being very dark or almost black, with indistinct markings. The richly tinted skins are highly valued, and are exported to Europe in large numbers, where they are used by the military officers for saddle coverings.

For capturing the Jaguar in steel-traps the directions given for trapping the cougar should be followed.

THE LION.

The principal habitat of the Lion is in Africa. Some also exist in Asia, but nowhere else. There are three African varieties — the Black, the Red or Tawny, and the Gray. In Asia the dark-colored Bengal, the light-colored Persian or Arabian, and the Maneless Lions exist. A full-grown Lion, in its native wilds, is usually four feet in height at the shoulders, and about eleven feet long from the nose to the tip of the tail. He is of great strength and ferocity, and is commonly called the "king of beasts." Lions belong to the cat family, and prey upon all animals they can master. They approach their prey stealthily, like a cat hunting a mouse, and spring upon it unawares. Human beings are not exempt from their attack, but form their most coveted prey when once an appetite for human flesh has been established. In Africa they hang round the villages, and carry off every man, woman, or child they can secure, and make great havoc among all kinds of domestic animals. Gérard, the French Lion-hunter of North Africa, estimates that the average length of life of the Lion is thirty-five to forty years; and that he kills, or consumes, year by year, horses, mules, horned cattle, camels, and sheep, to the value of twelve hundred dollars. Taking the average of his life, which is thirty-five years, each Lion costs the Arabs of that country forty-two thousand dollars. The Lion is mostly nocturnal in its habits, hunting its prey and satisfying its appetite during the night, and sleeping and digesting its food during the day. The Lioness is smaller than the male, and brings forth from one to three young at a time, about the beginning of the year. Lions are not numerous in Asia, and are steadily growing less so in Africa. They are now seldom found near the coasts of that Continent. Wherever the white man appears he wages relentless warfare against the "king of beasts." Its favorite haunts are the plains rather than the forests, and it is content with the shelter of a few bushes or low jungle. They sometimes hunt in troops — several attacking a herd of zebras, or other animals, in concert. Their strength is very great, and one has been known to carry a horse a distance of a mile from where he had killed it. Their most common prey are the deer and antelope which abound on the plains of Africa and in India. The zebra, the quagga, and the buffalo are their frequent victims.

The directions already given for taking the cougar with the steel-trap are adapted to the Lion. It may also be taken by setting a trap near its haunts and baiting it with a dead sheep or other animal. Great care must be taken to thoroughly secrete the trap, as the Lion is a very suspicious and intelligent beast. It is said that when a Lion is killed, all others retire from and avoid that immediate vicinity. The Lion is not a fastidious feeder. While, on the one hand, he likes to strike down a living animal and suck the hot blood from its body, on the other, he will devour any dead animal he may find, whether fresh or otherwise. "So thoroughly is this the case," says Wood, "that Lion-hunters are in the habit of decoying their mighty game by means of dead antelopes or oxen, which they lay near some water-spring, knowing well that the Lions are sure to seize so excellent an opportunity of satisfying at the same time the kindred appetites of thirst and hunger."

THE TIGER.

If the lion is the scourge of Africa, the Tiger holds that place in India and Southern Asia. The Royal Tiger of India rivals the lion in size, strength, ferocity, and activity, and excels him in beauty of form and color, and grace of movement. The Tiger is of great size, measuring in the largest specimens, four feet in height, four feet eight inches in girth, and thirteen feet six inches in total length. Its color is a tawny yellow, with transverse, dark-colored or black stripes. The under parts, the chest and throat, and the long tufts of hair on each side of the face are nearly white, and the markings on these parts are indistinct. The general make of the Tiger is a little more slender than that of the lion. Their haunts are the forests and jungles, and they prey upon all animals which come within their reach and power. They are of amazing strength and often bound upon their prey by a single leap of fifty feet. The Indian buffalo, which is as large as an ox, is killed and dragged off by the Tiger without difficulty. The female has from three to five young at a birth, which she defends with great fierceness. The range of the Tiger is confined to Asia, and to certain districts of that Continent. Some sections are terribly infested with them, and the inhabitants are kept in a state of terror by their depredations. They are common in the wilds of Hindostan, in various parts of Central Asia, even as far north as the Amoor River, and are also found on some of the large Asiatic Islands. Portions of Sumatra are so infested with them as to be almost depopulated. Here and in some parts of India, the Tiger is protected by the superstition of the people, who regard it as a sacred animal, animated by the souls of their dead ancestors, and none are killed but the " Man-eaters."

Wood in his Natural History gives the following description of the habits of the Tiger: —

" When seeking its prey, it never appears to employ openly that active strength which would seem so sure to attain its end, but creeps stealthily towards the object, availing itself of every cover, until it can spring upon the destined victim. Like the lion, it has often been known to stalk an unconscious animal, crawling after it as it moves along, and following its steps in hopes of gaining a nearer approach. It has even been known to stalk human beings in this fashion, the Tiger in question being one of those terrible animals called 'Man-eaters,' on account of their destructive propensities. It is said that there is an outward change caused in the Tiger by the indulgence of this man-slaying habit, and that a ' Man-eater' can be distinguished from any other Tiger by the darker tint of the skin, and a redness in the cornea of the eyes. Not even the Man-eating Tiger dares an open assault, but crawls insidiously towards his prey, preferring, as does the lion, the defenceless women and children as the object of attack, and leaving alone the men, who are seldom without arms.

"The Tiger is very clever in selecting spots from whence it can watch the approach of its intended prey, itself being couched under the shade of foliage or behind the screen of some friendly rock. It is fond of lying in wait by the side of moderately frequented roads, more particularly choosing those spots where the shade is the deepest, and where water may be found at hand wherewith to quench the thirst that it always feels when consuming its prey. From such a point of vantage it will leap with terrible effect, seldom making above a single spring and, as a rule, always being felt before it is seen or heard.

"It is a curious fact that the Tiger generally takes up his post on the side of the road which is opposite his lair, so that he has no need to turn and drag his prey across the road, but proceeds forward with his acquisition to his den. Should the Tiger miss his leap, he generally seems bewildered and ashamed of himself, and instead of returning to the spot, for a second attempt, sneaks off discomfited from the scene of his humiliation. The spots where there is most danger of meeting a Tiger, are the crossings of nullahs, or the deep ravines through which the water-courses run. In these localities the Tiger is sure to find his two essentials, cover and water. So apathetic are the natives, and so audacious are the Tigers, that at some of these crossings a man or a bullock may be carried off daily, and yet no steps will be taken to avert the danger, with the exception of a few amulets suspended about the person. Sometimes the Tigers seem to take a panic, and make a general emigration, leaving, without any apparent reason, the spots which they had long infested, and making a sudden appearance in some locality where they had but seldom before been seen. . . . . . .

"There is a certain bushy shrub, called the korinda, which is specially affected by the Tigers on account of the admirable cover which its branches afford. It does not grow to any great height, but its branches are thickly leaved, and droop over in such a manner that they form a dark arch of foliage, under which the animal may creep, and so lie hidden from prying eyes, and guarded from the unwelcome light and heat of the noonday sun. So fond are the Tigers of this mode of concealment that the hunters always direct their steps to the korinda-bush, knowing well that if a Tiger should be in the neighborhood, it would be tolerably certain to be lying under the sombre shade of the korinda branches."

There are a number of modes adopted by the natives of Asia, for killing the Tiger, such as spring-bows armed with poisoned arrows, nets, cages with trap-doors, enticing them into locations where they can be shot, &c.; but they are all inferior to the steel-trap. This instrument should be introduced wherever this lurking marauder abounds. The habit of returning to the unfinished carcass of the beast it has slain or found, which I have already noticed as pertaining to the cat family, is very strong in the Tiger, and can be taken advantage of in trapping them, in the same manner as described for the lion and cougar. The trap should be set near the hind parts of the carcass, as the Tiger always begins with those parts and eats toward the head. They may also be taken by setting traps along the paths which they make through the jungle near their lairs. In all cases the traps should be carefully secreted. A Tiger is easily killed with a bullet. Next to the brain and heart, the lungs and liver are its most mortal parts. A Tiger when struck by a bullet in the liver generally dies within fifteen or twenty minutes. If once wounded *anywhere* they usually die, though perhaps not immediately. From some unknown cause a wound on a Tiger very soon assumes an angry appearance, becomes tainted and the abode of maggots, and finally proves fatal. This tendency to putrefaction in the Tiger, renders it necessary that they should be skinned immediately after they are killed if the preservation of the skin is any object. Especially should the Tiger be removed out of the sunshine, instantly after it is slain. A delay of ten or fifteen minutes will often ruin the skin by the loosening of the hair from putrefaction. The skin after being removed should be at once stretched, and treated with a very strong solution of salt, alum, and catechu.

Several other large animals of the cat kind are found in Asia and Africa, such as the Leopard, the Ounce, the Riman-

Dihan or Tree-Tiger, &c. They are all carnivorous and of similar habits, and should be trapped on the same general principles as the tiger and cougar. Of these animals, the Leopard is the most formidable and destructive. It is found in both Asia and Africa, but in greatest numbers in the latter country. It is much smaller than the tiger, but of extraordinary strength for its size. It does not usually attack man, unless wounded or pursued. It is very destructive among the herds of domestic animals, antelope, deer, and monkeys. It is celebrated for the beauty of its skin and the agility and grace of its movements. Its haunts are the forests where thick, high undergrowth prevails.

THE WOLVERENE.

This animal is found throughout a large part of British America, and in some of the wildest portions of the Northern States. It is about three feet long from the nose to the root of the tail, and has a tail fourteen inches in length. In general appearance and movements it resembles the bear, while its head bears a strong likeness to that of the fisher except that the muzzle is shorter. The habits and food of the Wolverene are much like those of the marten. They hunt hares, mice, birds, and kill disabled deer. They are powerfully built and possess great strength. Their prevailing color is dark brown on the back and under parts. A broad stripe of yellowish brown sweeps along each side and ends at the root of the tail. The legs and feet are black. Stripes and patches of black and yellow occur on the under parts. The fur is long, soft, and tolerably fine, overlaid with larger and coarser hairs, which are about three inches long on the rump but shorter in front. The Wolverene is a great mischief-maker for the trapper in the regions where it dwells, especially the marten-trappers of British America, who use the old-fashioned "dead-fall." One of these animals will follow a line of traps for miles, tearing them down, devouring bait and the animals that have been caught. They are also very troublesome in destroying *caches* of provisions. On account of its destructive propensities, and great cunning and sagacity, the Indians

THE VIRGINIA OPOSSUM.

call the Wolverene the Evil One or Devil. They are seldom caught in traps, and the most successful way of destroying them is said to be by strychnine.

THE OPOSSUM.

This animal inhabits the warmer parts of the United States, and several species of it are said to exist also in Australia. In form it somewhat resembles the common house rat. Its body is about twenty inches long, stoutly built, and its tail, which is generally fifteen inches in length, is prehensile, like that of some monkeys, i. e., capable of holding on to any thing that it encircles. The Opossum is five-toed, and walks on the sole of its foot like the bear. Its ears are large, rounded, and almost naked. The female has from nine to thirteen teats, the odd one being in the centre of the ring formed by the rest. The fur is long, soft, and woolly, whitish at the roots, and brown at the top. The Opossum is omnivorous, feeding on corn, nuts, berries, roots, insects, young birds, eggs, mice, &c. It is nocturnal in its habits; hiding in the thick foliage of the trees in the daytime, and seeking its food by night. It is an active climber, and is said to spend much of its time and even to sleep suspended from the limb of a tree by the tail! The females are very prolific, producing from nine to thirteen young at a birth, and three or even four litters in a year. They are provided with a pouch under the belly, in which they protect and suckle their young.

These animals are trapped in the same manner as the raccoon and the badger, by setting traps in their haunts, and baiting with any of their favorite kinds of food. They have a habit, when caught, of feigning death, and will bear considerable torture without betraying any signs of life. This habit doubtless gave rise to the common by-word which calls certain kinds of deceit "playing 'possum."

THE SKUNK.

This animal, though generally much despised in this country, is said to furnish the staple fur to Poland, and deserves at least the respectful attention of the trapper. It is related to the weasel. Its head is small, with a projecting, naked nose, small, piercing eyes, and short, rounded ears. The body is about eighteen inches long; the tail twelve or fourteen inches, and bushy. The feet are short, and well adapted to digging, having naked soles and closely united toes with claws. The prevailing color is white and black, some varieties being mostly white and others mostly black. The fur of the latter is the most valuable. The Skunk walks with its back much curved, and its tail erect, as though proud of its beauty. It is nocturnal in its habits, and during the summer months searches the fields in the vicinity of its haunts every night, feeding principally on worms, bugs, and grasshoppers, but sometimes devouring frogs, mice, young birds, green corn, &c., and occasionally making free with poultry and eggs. Its services in clearing the farmer's fields and gardens of bugs and worms more than pay for its depredations, and it ought to be regarded as a useful animal. Its breeding season is in April or May. From six to nine young are brought forth at a litter, and are reared in holes or among rocks, till they are large enough to shift for themselves.

These animals are taken in traps set at the mouths of their holes or in the fields where they search for food. The trap should be covered with loose earth or soft vegetable substances, and should be baited with small pieces of meat scattered around it. They are not cunning, and require no great skill in taking them. The great difficulty in trapping for them or meddling with them in any way is in the liability of catching a charge of their perfumery, which is very disagreeable, and ruins all clothing that is once impregnated with it. This offensive essence is ejected from two glands near the anus by the contraction of the muscular coverings, and the only way that I know to prevent the discharge is to approach the animal in the trap stealthily, and give it a smart blow with a club across the back near the tail, which will paralyze the ejecting muscles. But this expedient is not always available, as the animal sometimes takes the trap for a living enemy and discharges when first taken. One thing, however, is in its favor, namely, it is very neat in its personal habits, rarely allowing

The Skunk.

its own fur to be soiled with its offensive secretions; so that if you can get away its skin without being overwhelmed yourself by its perfumery, your spoil is likely to be as clean and saleable as in the case of any other animal.

[We are indebted to an old Connecticut trapper, Mr. H. Mansfield, for the following valuable addition to Mr. Newhouse's article on the Skunk. — EDITORS.]

"In summer, Skunks can be taken in great numbers by the following method: Find a place where they travel from their holes to a hen-coop or through a corn-field. Make a path for them by treading down the grass, and set up sticks along on each side to guide them more surely. Set traps at intervals, and strew pieces of meat or dead mice before and behind each trap. A whole family of Skunks will walk down this path, the old ones heading the procession; and as one after another is caught, those behind will climb over and pass on, till all are taken. I have caught in this way two old ones and eight young ones in one path on a single evening. They seldom discharge when first caught; and can be prevented from doing so at all, either by a blow on the back, or by boldly seizing the parts where the offensive secretion lies with one hand, and piercing the throat with a knife in the other.

"In winter my method is to track them to their holes and dig them out. They are obliged to go to some stream for water every day, and when there is snow, they can easily be tracked back to their burrows. In digging them out, I prevent them from using their terrible weapon by carefully uncovering only one at a time, and only the head of each at first, filling in and even 'tamping' the dirt around the body, till I can despatch them in succession by opening the jugular vein.

"The surest way to take Skunks without bad consequences is by the snare and spring-pole.

"With all the precaution that can be taken, the trapper's clothes will sometimes be sprinkled; and there will be more or less scent about the skins. The best way to cleanse articles in this condition is to hold them over a fire of red-cedar boughs, and afterwards sprinkle them with chloride of lime."

THE COYPU RAT.

The Coypu Rat, or Racoonda, as it is sometimes called, furnishes the fur known in commerce as Nutria. But one species is known, which is a native of South America, and is found in great numbers in the La Plata region. In general appearance and character it resembles the beaver. Its tail, however, instead of being flattened, is long, round, and rat-like. Its favorite haunts are the lagoons of the plains or pampas, and the banks of rivers and streams. Its fur is short, fine, silky, similar to that of the beaver, and light brown in color. Overlying the fur are long hairs of a brownish yellow color. The fur is heaviest and best on the belly. It is used for the same purpose as that of the beaver, in the manufacture of hats and caps. The Coypu is about two feet long exclusive of tail, which is about fifteen inches in length. It is very prolific, the female producing six or seven at a birth. They feed on vegetables, are quite gentle in their character, and easily tamed. They inhabit South America on both sides of the Andes: on the east, from Peru to forty-three degrees south latitude; on the west, from Central Chili to Terra del Fuego. They are also found in the small bays and channels of the archipelagos along the coast. They are burrowing animals, and form their habitations in the banks of lakes and streams. They are nocturnal in their habits, and seem to be equally at home in fresh or salt water.

The Coypu is usually hunted with dogs, and is easily captured. It is, however, a bold animal, and fights fiercely with the dog employed in pursuing it. We cannot learn that any attempt has been made to take them by the steel-trap, but this would no doubt prove the best and easiest method of capture. Their habits resemble those of the beaver and muskrat, and they should be trapped on the same general principles. Great numbers of the skins of this animal are annually exported. In some seasons the number has been over three millions, constituting an important branch of the fur-trade.

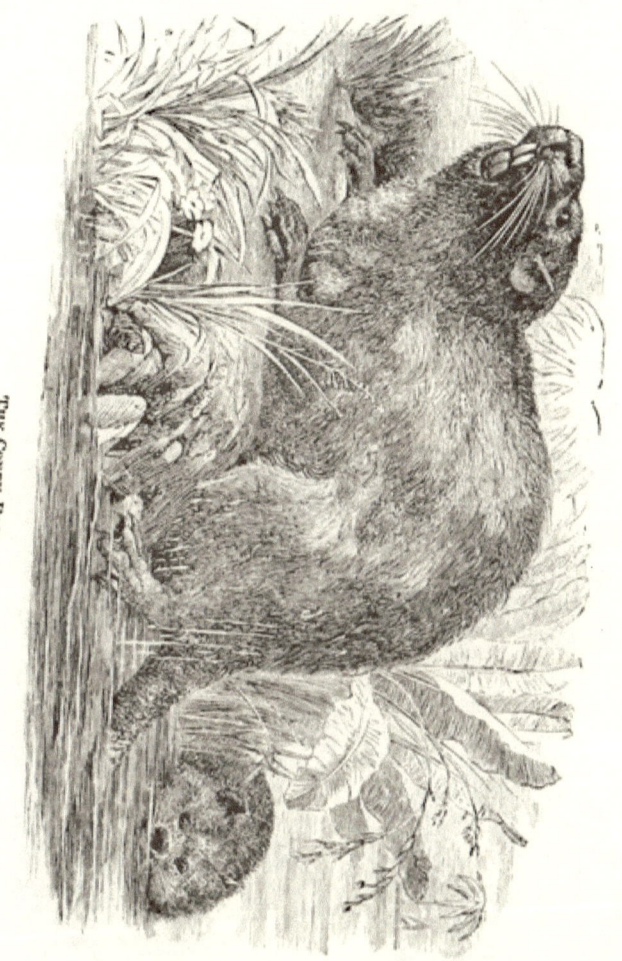

The Coypu Rat.

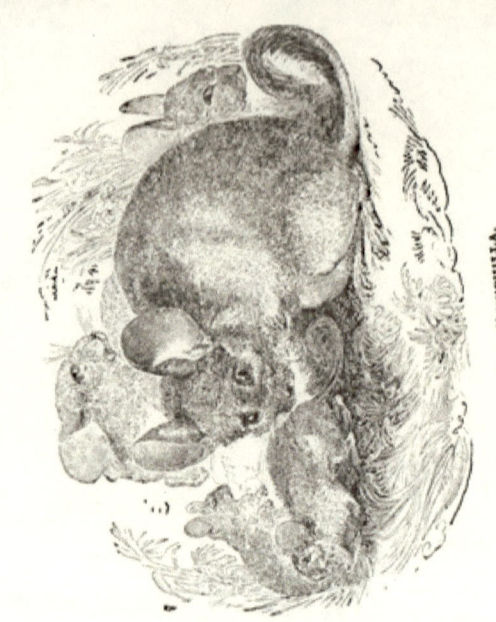

THE CHINCHILLA.

THE CHINCHILLA.

The most delicate and silken of all furs is that produced by the Chinchilla. This animal is found in South America, along the Andean region from Chili to Peru. It burrows in the valleys which intersect the hilly slopes, and collects together in great numbers in certain favored localities. It belongs to the group of animals called *Jerboidæ*, which are characterized by great comparative length of the hind legs. It is a small animal, measuring only about fourteen or fifteen inches in total length, of which the tail forms about one third. They are very prolific, the female bringing forth five or six twice a year. Their food is exclusively vegetable, consisting mostly of bulbous roots. They are very cleanly in all their habits. The fur of the Chinchilla is long; its color is a delicate clear gray upon the back, softening into a grayish white on the under portions; and its texture is wonderfully soft and fine. It is used for muffs, tippets, linings to cloaks and pelisses, and trimmings. The skins which are obtained in Chili are the best. Great numbers of Chinchillas are caught in the vicinity of Coquimbo and Copiapo. They are usually hunted with dogs by boys. The true method is to take them at the mouth of their burrows with a small steel rat-trap.

THE SQUIRREL.

The American varieties of the Squirrel do not produce fur of much value, and are of little importance in the fur-trade. They are generally taken only for food or as nuisances. The European variety, however, is much more valuable, and its skins are brought into the fur-markets of Europe by the million. They are spread over all the north of Europe and Asia. Those of Russia and Siberia produce the finest and handsomest fur. This kind is a small Squirrel with tufted ears and a beautiful gray coat.

For taking Squirrels, the trap should be set on the top rail of a fence near a wood that they frequent. A pole, with an ear of corn or some other favorite squirrel-food fastened to the end of it, should be set up by the side of the fence, leaning in such a position as to bring the bait over the trap at the height of six or eight inches. In reaching for the bait the Squirrel gets into the trap.

To give a complete view of the business of trapping, several less valuable animals should be briefly noticed, not as fur-bearing, but as legitimate subjects of the trapper's art.

THE WOODCHUCK OR MARMOT.

Marmots are burrowing animals. There are a number of species, and they are found on both Continents. In this country, they are commonly called Woodchucks. The curious Prairie Dog of the Western plains is allied to the Marmot. This latter animal lives in villages from a few acres to several miles in extent, in the country bordering on the Arkansas and Missouri Rivers and their tributaries. The entrance to their burrows is in the summit or side of a small mound of earth, somewhat elevated, but seldom more than eighteen inches high. In pleasant weather, they may be seen sporting about the entrance of their burrows; and five or six individuals may be sometimes seen sitting on a single mound. They make a noise somewhat like the barking of a dog, whence their name, Prairie Dog. When alarmed, they retreat at once into their holes. The skin of the common Woodchuck is valuable for whip-lashes, and its fur even is not despised by rustics. All kinds of Marmots may be taken by setting steel-traps, completely covered and without bait, at the mouth of their holes.

THE GOPHER.

This animal, called also the Canada Pouched Rat, inhabits the prairie region west of the Mississippi. It is a burrowing animal, and lives on roots and vegetables. Its body is firmly built, about nine inches long, with a short tail and legs, the latter armed with long claws for digging. The head and neck are relatively large, and the mouth has four broad long incisors, two on each jaw, adapted to cutting roots. On the sides of the face and neck, extending back to the shoulders,

The Woodchuck, or Marmot.

are large pouches, in which to carry earth, food, &c. The Gopher digs paths or galleries of an oval form, several inches in diameter, a short distance below the surface, coming to the surface once in about a rod, where the excavated earth is deposited in little hillocks. These galleries ramify in all directions. When the animal has brought to the surface in one place as much earth as its sense of economy dictates, it closes up the hole, and begins a new deposit further on, so that noth-

The Gopher and its Burrow.

ing remains but a neat little mound of earth, large enough to fill a half bushel, more or less. Gophers are great pests to the western farmers, injuring and destroying the roots of their crops, and infesting their fields and gardens. They may be trapped in the following manner: Carefully cut away a square section of sod on a line between the two most recent deposits. On finding the gallery, excavate down till a trap will set on a level with the bottom of the passage. Place the trap there; then lay a piece of board or shingle across the excavation, just above the passage, and replace the sod. The Gopher while at work will run into the trap and be taken.

THE RAT.

This pest of all countries may be taken in any or all of the following ways : 1. Set your trap in a pan of meal or bran; cover it with meal and set the pan near the run-ways of the Rats ; or, 2, set the trap in a path at the mouth of a Rat's hole, with a piece of thin brown paper or cloth spread smoothly over it; or, 3, make a run-way for the Rats by placing a box, barrel, or board near a wall, leaving room for them to pass, and set the trap in the passage, covered as before. In all cases, the trap should be thoroughly smoked over a fire or heated over a stove before it is set, and at every re-setting; but care should be taken not to overheat the trap so as to draw the temper of the spring. Also the position of the trap should be frequently changed.

To conclude these instructions for capturing animals, I will introduce the trapper to one or two of a larger and nobler family, which he will find well worthy of his attention, not for their skins or furs (though these are valuable), but for their flesh, which, in his more distant and adventurous excursions, will often be the only resource of his commissariat. The soldier must look out, not only for his means of fighting, but for his means of living — for his larder as well as for his enemy — and happily I can show the soldiers of the trap how to supply themselves with food by the same weapons that they use in taking animals for their furs.

THE DEER.

This family of ruminating animals embraces a great variety of species, ranging in size from the Pigmy Musk-Deer of Java, which is not larger than a hare and weighs only five or six pounds, to the gigantic Moose-Deer of America, whose height is seven or eight feet and its weight twelve hundred pounds. But the species with which American trappers are most practically concerned, are the common Red or Virginia Deer, and

The Rat.

the Black-Tailed Deer of the region west of the Mississippi. These species differ but little in habits and general characteristics, and a description of the Virginia Deer is sufficient for the purposes of the trapper. The Virginia Deer are found in nearly all the States of the Union east of the Rocky Mountains, and abound in both provinces of Canada. They are gregarious in their habits, though frequently seen alone. Their food in summer consists of twigs, grass, berries, nuts, roots, acorns, persimmons, &c., and at that season they frequent rivers and lakes to feed on water-plants, as well as for the purpose of freeing themselves from insect pests. They are also fond of visiting the pioneer's clearing and appropriating his wheat, corn, oats, potatoes, turnips and cabbages. In winter they retire to the elevated ridges, where maple and other hard-wood trees abound, the bark, twigs and branches of which are at that season their chief support. They form "yards" by trampling down the deep snows, and live together in large herds, numbering sometimes thirty animals in a single "yard." These inclosures are enlarged from time to time as the Deer require more trees for browsing. Wolves and panthers are their most formidable enemies — always excepting man. Packs of wolves frequently attack them in their "yards," and sometimes when the snow is deep and crusted over, whole herds are destroyed. Wolves sometimes pursue a single Deer with the "long chase." In summer a Deer thus pursued generally takes to the water, and so baffles his pursuers; but in winter when the streams and lakes are frozen over, he rarely escapes. Panthers take Deer by crawling within springing distance of them in their "yards" or elsewhere, or by watching and pouncing on them from some cliff or tree, as they pass below.

The methods by which men take Deer are various. They are sometimes driven by dogs into rivers or lakes, and are then overtaken and dispatched by the hunter in his canoe. A favorite method is to shoot them at night at the places by the water-side, where they resort to feed on aquatic plants and relieve themselves of insects. For this purpose the hunter prepares himself with a boat, gun, and lamp. The light is set on the bow of the boat, so that it will shine on the forward sight of the gun, and at the same time conceal by its glare the hunter crouching behind. With muffled oar the boat approaches the game. The reflected gleams from the eyes of the Deer betray his position to the hunter. If no noise is made the victim will stand and gaze at the light until it is within a few yards, and so give a sure opportunity for the fatal shot. Many are taken in this way in the early autumn; and later in the season, when snow first comes, many more are taken by the "still hunt," either by following on their trail, or by watching at their run-ways.

The steel-trap, it must be confessed, is not much used for taking Deer; and I am not sure but that this use of it is regarded by sportsmen as somewhat barbarous. But all the ways of deceiving and killing these noble animals seem to be open to the same objection; and the necessities of the trapper often forbid him to be very particular as to the means of furnishing himself with food. There are times when the trap is the best, and even the only, available means of taking Deer; for instance, when the trapper is without his rifle, or has exhausted his ammunition, and finds himself in the far wilderness without food. In such circumstances, he might starve if he could not betake himself to his traps for supply. And even when rifle and ammunition are at hand, sometimes in dry weather (technically called a "noisy time") every thing is so crisp and crackling under foot, that it is impossible to approach the Deer within shooting distance. I therefore recommend to practical woodsmen to learn how to take Deer in traps, and not be over-scrupulous in doing so when occasion requires.

For taking Deer the trap must be a strong one, and the jaws should be spiked, and so shaped and adjusted that when sprung they will remain open about half an inch, to prevent breaking the bone.

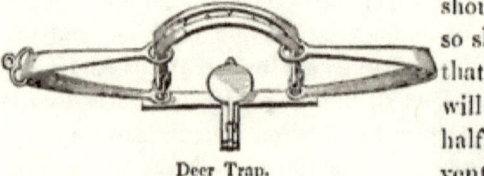

Deer Trap.

The trap should be placed in the path of the deer where it

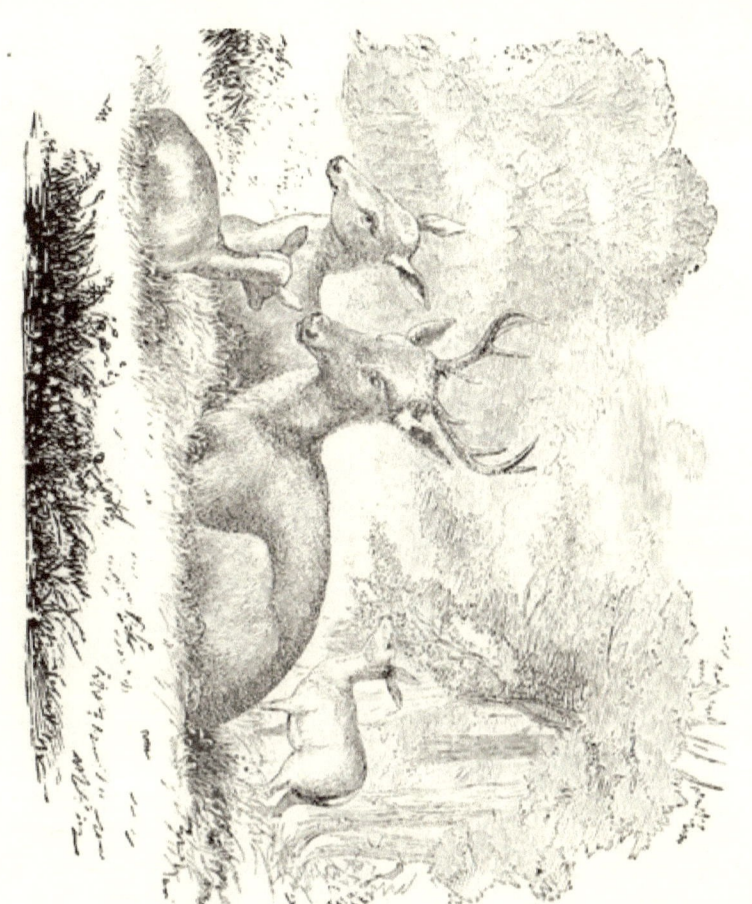

FAMILY OF DEER.

crosses a stream or enters a lake; and it should be set under water and concealed by some covering. If it is as heavy as it ought to be (say of three or four pounds' weight), it should not be fastened at all or even clogged; as the animal is very active and violent when taken, and will be sure to break loose by tearing off a limb or smashing the trap, if his motions are much impeded. If the trap is left loose, the Deer, when caught, will make a few desperate plunges and then lie down; and will seldom be found more than ten or fifteen rods from where he was taken. When the hunter approaches he will make a few more plunges, but can easily be dispatched.

Mr. Gunter, the Canada trapper, whom I have heretofore quoted, gives the following directions for trapping Deer in winter: —

"Fell a maple or bass-wood tree near where the Deer haunt. These trees furnish their favorite browse. Make a small hole in the snow, close to the top of the tree. Set your trap, lower it into the hole and shove it to one side, eighteen or twenty inches, through the snow. Finally take some deer-scent, obtained from the glands on the hind legs of a Deer, and which has a very strong odor, and rub it on your trap. This done, when the Deer come to feed on the twigs of the fallen tree, you will be pretty sure to take one."

THE MOOSE.

This is the largest kind of deer, and its habits are in many respects like those of the common deer. It is more confined, however, to the snowy regions of the North. It is found throughout the greater part of British America, ranging as far north as the Arctic Sea. In the United States, it is found in Maine, Northern New York, Oregon, and Washington Territory. On the Eastern Continent, it is found throughout the northern parts of Europe and Asia. Its favorite haunts are the hard-wood lands. In general color, it is yellowish-brown or ashy-gray. The hair in summer is short and soft, and long and coarse in winter. The full-grown Moose weighs from eight hundred to fifteen hundred pounds, and stands seven and even eight feet high. Its horns have an expanse of nearly six feet between the tips, and a palm or spade on each, of a foot in width, and weigh from forty-five to seventy pounds. Under the throat of both sexes there is a tuft of coarse, bristly hair, a foot or more in length, attached to a sort of dewlap. The breeding season of the Moose is in May. At the first birth, but a single one is brought forth; afterwards two are brought forth annually. Moose, like the common deer, frequent rivers and lakes in summer, to feed on the roots of the water-lily and other aquatic plants; and retire in winter to the high ridges, to browse on the twigs of the striped maple and birch. Their height enables them to crop the overhanging branches of large trees; and their weight and strength enable them to bend down small trees and slide over them with their bodies, stripping the bark and twigs to the very extremities. Like the common deer, they form "yards" by treading down the snows, and enlarge them as fast as they strip the trees and require more. In these "yards" there are commonly found a male, female, and two fawns.

Moose are taken in winter by the "long chase" on snow-shoes, and in summer they are shot at their feeding-places in marshes. They are, however, very wary and timid; and their sense of smelling is so acute that the greatest caution is necessary on the part of the hunter in approaching them. The males in the rutting season are very dangerous, and will attack, and if possible kill, any persons who approach them. Moose can easily be taken either in summer or winter by setting steel-traps in their haunts, as they are not cunning, and enter a trap as readily as an ox or a horse. The trap should be a strong one of about forty pounds' weight, and it should be fastened to a clog of sixty pounds' weight.

The flesh of the Moose is much esteemed by hunters and trappers, being generally preferred to that of the common deer. The marrow in the large bones is an excellent substitute for butter.

MOOSE-YARD.

III. CURING SKINS.

However successful a trapper may be in taking animals, he will not secure a full reward for his labor unless he knows how to take care of their skins, and prepare them for market in such a manner that they will command the highest prices. As skins that have been riddled with shot find little favor with fur-dealers, so skins that have been cut in stripping off, or that are encumbered with remnants of flesh, or that have passed into a state of incipient putrefaction before drying, or that have not been properly stretched, or that have been dried too fast, or that have been neglected and exposed after being cured, are very sure to be thrown out by the fur-inspector as second or third rate skins, deserving only poor prices. Great quantities of valuable furs, taken by boys and inexperienced trappers, are rendered almost worthless by bad treatment in some of the processes of preservation. I shall give such information on this part of the trapper's business as I have obtained, both from my own experience and from conversation with fur-dealers.

GENERAL RULES.

1. Be careful to visit your traps often enough, so that the skins will not have time to get tainted.

2. As soon as possible after an animal is dead and dry, attend to the skinning and curing.

3. Scrape off all superfluous flesh and fat, but be careful not to go so deep as to cut the fibre of the skin.

4. Never dry a skin by the fire or in the sun, but in a cool, shady place, sheltered from rain. If you use a barn door for a stretcher (as boys sometimes do), nail the skin on the *inside* of the door.

5. Never use "preparations" of any kind in curing skins, nor even wash them in water, but simply stretch and dry them as they are taken from the animal.

STRETCHING SKINS.

In drying skins, it is important that they should be stretched tight, like a strained drum-head. This can be done after a fashion by simply nailing them flat on a wide board or a barn door. But this method, besides being impracticable on the large scale in the woods (where most skins have to be cured), is objectionable because it exposes only one side of the pelt to the air. The stretchers that are generally approved and used by good trappers are of three kinds, adapted to the skins of different classes of animals. I shall call them the *board-stretcher*, the *bow-stretcher*, and the *hoop-stretcher*, and will describe them, indicating the different animals to which each is adapted.

THE BOARD-STRETCHER.

This contrivance is made in the following manner: Prepare a board of bass-wood or other light material, two feet three inches long, three inches and a half wide at one end, and two inches and an eighth at the other, and three eighths of an inch thick. Chamfer it from the centre to the sides almost to an edge. Round and chamfer the small end about an inch up on the sides. Split this board through the centre with a knife or saw. Finally, prepare a wedge of the same length and thickness, one inch wide at the large end, and tapering to three eighths of an inch at the small end, to be driven between the halves of the board. This is a stretcher suitable for a mink or a marten. Two larger sizes, with similar proportions, are required for the larger animals. The largest size, suitable for the full-grown otter or wolf, should be five feet and a half long, seven inches wide at the large end when fully spread by the wedge, and six inches at the small end.

Board-Stretcher. An intermediate size is required for the fisher, raccoon, fox, and some other animals, the proportions of which can be easily figured out.

These stretchers require that the skin of the animal should not be ripped through the belly, but should be stripped off whole. This is done in the following manner: Commence with the knife at the hind-feet, and slit down to the vent. Cut around the vent, and strip the skin from the bone of the tail with the help of the thumb-nail or a split stick. Make no other slits in the skin, except in the case of the otter, whose tail requires to be split, spread, and tacked on to the board. Peel the skin from the body by drawing it over itself, leaving the fur-side inward. In this condition the skin should be drawn on to the split board (with the back on one side and the belly on the other) to its utmost length, and fastened with tacks or by notches cut in the edge of the board, and then the wedge should be driven between the two halves. Finally, make all fast by a tack at the root of the tail, and another on the opposite side. The skin is then stretched to its utmost capacity, as a boot-leg is stretched by the shoemaker's " tree," and it may be hung away in the proper place, by a hole in one end of the stretcher, and left to dry.

A modification of this kind of stretcher, often used in curing the skins of the muskrat and other small animals, is a simple board, without split or wedge, three sixteenths of an inch thick, twenty inches long, six inches wide at the large end, and tapering to five and a half inches at six inches from the small end, chamfered and rounded as in the other cases.

Muskrat-Stretcher.

The animal should be skinned as before directed, and the skin drawn tightly on to the board, and fastened with about four tacks. Sets of these boards, sufficient for a muskrat campaign, can easily be made and transported. They are very light and take up but little room in packing, thirty-two of them making but six inches in thickness.

THE BOW-STRETCHER.

The most common way of treating the muskrat is to cut off its feet with a hatchet; and rip with a knife from between the two teeth in the lower jaw, down the belly, about two inches below where the fore-legs come out. Then the skin is started by cutting around the lips, eyes, and ears, and is stripped over the body, with the fur-side inward. Finally a stick of birch, water-beech, iron-wood, hickory, or elm, an inch in diameter at the butt, and three feet and a half long, is bent into the shape of an ox-bow and shoved into the skin, which is drawn tight, and fastened by splitting down a sliver in the bow, and drawing the skin of the lip into it.

This method is too common to be easily abolished, and is tolerable when circumstances make it necessary; but the former method of stretching by a tapering board, in the case of muskrats as well as other small animals, is much the best. Skins treated in that way keep their proper shape, and pack better than those stretched on bows, and in the long run boards are more economical than bows, as a set of them can be used many times, and will last several years; whereas bows are seldom used more than once, being generally broken in taking out.

THE HOOP-STRETCHER.

The skins of large animals, such as the beaver and the bear, are best dried by spreading them, at full size, in a hoop. For this purpose, a stick of hickory or other flexible wood should be cut, long enough to entirely surround the skin when bent. (If a single stick long enough is not at hand, two smaller ones can be spliced together.) The ends should be brought around, lapped, and tied with a string or a withe of bark. The skin should be taken from the animal by ripping from the lower front teeth to the vent, and peeling around the lips, eyes, and ears, but without ripping up the legs. It should then be placed inside the hoop and fastened at opposite sides, with twine or bark, till all loose parts are taken up, and the whole stretched so that it is nearly round and as tight as a drum-head. When it is dry it may be taken from the hoop, and is ready for packing and transportation.

This is the proper method of treating the skin of the deer. Some prefer it for the wolf and raccoon. In many cases the trapper may take his choice between the hoop and the board method. One or the other of these methods will be found satisfactory for curing all kinds of skins.

IV. LIFE IN THE WOODS.

[The outfit for campaigning in the woods proposed by Mr. Newhouse in the following chapter may seem rather elaborate and luxurious, adapted perhaps better to amateur sportsmen than to the "rough and ready" followers of the trap. But it is best to encourage and help forward as far as possible good civilized living, even in wild places. Those who prefer a freer and less expensive style of outfit can leave Mr. Newhouse and take lessons of the older trapper, John Hutchins, or of Mr. Gunter, both of whose programmes are given further on, and are simple enough for the hardiest. — EDITORS.]

THE great question, after all, for the trapper as well as for the soldier, is, how to live and keep himself comfortable while he carries on the war. He requires in some respects even more than a soldier's courage, for he is to encounter the hardships of camp-life *alone*, or with but one or two companions, and without a baggage-train to bring up provisions at every halt. The very first article of outfit that he should equip himself with, I should say, would be a firm trust in Providence. But as Cromwell told his soldiers to "trust God and keep their powder dry," so the trapper will need to provide some things for himself, while he trusts Providence. I will therefore tell him as well as I can, how I equip myself for life in the woods.

OUTFIT FOR A CAMPAIGN ON FOOT.

If the region in which you propose to trap cannot be reached by boat or wagon, you must be content with such necessaries as you can carry on your person. A trapper on foot should not tire himself with long stiff-legged boots, but should wear short half-boots (with soles well nailed), fitting snugly above and around the ankle. His pants should be gray woolen, closely fitting below the knee, but roomy above. His coat should be of the same material and color, with plenty of

MR. NEWHOUSE'S TENT AND STOVE.

pocket-room. His hat should be of soft felt, gray, and with a moderate brim. He should carry a "change" of woolen drawers, wrappers, shirts, and stockings. A towel with soap, a night-cap, and a blanket, or, what is better, a Canton-flannel bag to sleep in, will complete his personal equipments. Then he must carry for shelter a small tent, made of firm cotton-drilling, weighing not more than two pounds and a half;

Shelter Tent.

for subsistence, a double-barrelled gun (rifle and shot), weighing seven or eight pounds, with ammunition, and fishing-tackle; and, for all sorts of purposes, an axe of two and a half pounds (with a good length of handle), and plenty of tacks and nails. For cooking and table service he must carry a frying-pan, a camp-kettle, a hunting-knife, some knives and forks, spoons of two sizes, a few tin pressed plates and basins, and a drinking-cup. Above all, he must not forget to take a good supply of matches and a pocket-compass. These necessaries (exclusive of clothing) will weigh, according to my reckoning, about twenty-five pounds. The rest of his load must be made up of traps and provisions. If he is stout enough to undertake trapping on foot, he ought to be able to travel with about fifty pounds. He may take then five pounds of provisions and twenty pounds of traps, or any other proportion of these articles that will make up the remaining twenty-five pounds. His provisions should consist of articles that will be desirable as accompaniments to the produce of his gun and fishing-tackle, namely, sugar, tea and coffee (rather than whiskey), salt, pepper, butter, lard, sifted Indian meal, white beans, crackers, &c. The butter and lard should be put up in air-tight cans, and on arrival at the trapping grounds should be sunk in a spring. The best kind of knapsack for carrying such an outfit is made of rubber-cloth, with shoulder-straps; but you can easily convert your sleeping bag or your blanket into a knapsack that will serve very well.* If you trap with one companion or more (which is a good plan and according to the general practice), many of the articles named in the above list will answer for the party, and so the load for individuals will be lightened. Thus equipped, you can turn your back on the haunts of men, march into the wilderness, and, with a little hunting and fishing in the intervals of trap-duty, live pleasantly for months, and return with your load of furs, a stouter and healthier man than when you started.

OUTFIT FOR AN EXCURSION BY WAGON OR BOAT.

If your trapping district can be reached by road or by water, some changes should be made in the foregoing inventory. For the interest of your larder it will be best to take more ammunition, and a greater variety of fishing-tackle. A lamp and lantern, with a supply of oil, a camp-hatchet of twelve ounces in weight with a fourteen-inch handle, a stone for sharpening knives, axes, and hooks, a magazine of needles, thread, scissors, &c., and many other like conveniences, may

---

* One of the most satisfactory arrangements we have ever seen for carrying luggage on the back is the Indian shoulder-basket. They are made nearly square, or about ten inches by twelve, at the bottom, and twelve or fourteen inches high. One side is flat, the others are rounded and drawn in toward the top, making the mouth of the basket only about half the size of the bottom. Over the mouth, and extending some distance down the sides, a cover of rubber or enamel-cloth should be fitted. On the flat side of the basket shoulder-straps are fastened, crossing each other in the form of an X. These straps should be made of two thicknesses of strong cotton cloth, sewed together and stuffed with cotton. The great advantages of this basket are, that it is light, easily managed, fits the back well, bringing the load just where it is wanted, does not get out of place, and does not heat the back like a close-fitting knapsack. Combined with the basket the trapper needs a small enamel-cloth haversack such as is worn by soldiers. — EDITORS.

be stowed away in the odd corners of your luggage. You may also carry more clothing and more provisions, such as potatoes, and ought certainly to take along at least one hundred and fifty traps of different sizes, and a good set of board-stretchers for curing skins.

TENT.

In the place of the light half-tent recommended for a campaign on foot, you should take a regular A tent of eight or nine pounds' weight, house-shaped, and buttoning up in front. This should be dipped two or three times in a solution prepared by mixing equal parts of sugar of lead and alum in a pailful of milk-warm water. This treatment will render the tent almost impervious to rain, and will protect it from the sparks of fire that will occasionally be blown upon it. Instead of a ridge-pole and two forked stakes for supporting it, all you need is a cord thirty or forty feet long, to be drawn through the ridge of the tent, fastened to it about midway, and tied at the ends to two trees at the proper height. The sides should be drawn down tight and fastened by hooks driven into the ground.

STOVE AND FURNITURE.

A much needed convenience for life in the woods is a stove with its furniture, that shall on the one hand afford all necessary facilities for cooking and warming, and on the other shall take up the least possible room in packing. Having devoted considerable study to this matter, I flatter myself that I can put the ingenious trapper in a way to make or procure the exact article that he wants. Your stove should be made of sheet-iron, and should be twenty-seven inches long, ten inches wide, and eight inches deep, having on the top two eight-inch holes for boilers and one four-inch hole for the smoke-pipe. Ten feet of pipe will be sufficient, and this can be made in five joints of two feet each, tapering in the whole length from four inches in diameter to three, so that the joints will slip into each other and the whole can be packed for transportation inside the stove. For an outlet of the pipe through the roof of the tent, there should be a piece of tin, ten inches square, with an oblong hole, to be fastened at the proper place on the roof by means of lappels. The furniture of the stove should be two dripping-pans of Russia iron; one thirteen inches long, nine inches wide, and an inch and a quarter deep; the other enough smaller to pack inside the first; a kettle, also of Russia iron, nine inches across the top, seven inches and a half deep, and six inches and a half across the bottom; and two or three tin pails and several basins, all made in a diminishing series, so that they will slip into each other, and all into the iron kettle. The kettle and pails match the holes in the top of the stove, and when used in cooking tea, coffee, &c., should be covered with tin pressed plates. The whole of this furniture can be packed with the pipe in the stove. For supporting the stove in the tent, prepare four posts eighteen inches long, made of three-eighths inch iron rod, sharpened at one end, flattened at the other and fashioned like a small tenon. Two pieces of band-iron should then be made just long enough to reach across the bottom of the stove and receive the tenons of the posts into holes drilled in each end. Then, to set up your stove, drive the posts into the ground, adjust the cross-pieces to their places, and place the stove on the cross-pieces. Small depressions should be filed in the edge of the stove-bottom, to fit the ends of the tenons, above the cross-pieces, so as to prevent the stove from moving from its position. Your tent is large enough to accommodate any number of persons from two to six; and your stove will warm them and do their cooking, with an amount of fuel that will be a mere trifle compared with what is required for an open fire. It has the advantage also of giving a quick heat, and, with a damper, will keep fire all night.

BED AND BEDDING.

Good sleeping accommodations can be provided in the following manner: Take two pieces of sacking or other coarse cloth, six and a half feet long and two feet and three quarters wide, and sew them firmly together at the sides, making a bag with both ends open. Cut two poles, each seven feet long and two inches in diameter, and run them through the bag, resting the ends in notches on two logs placed parallel to each other at the proper distance apart. The notches should be so far apart that the poles will tightly stretch the bag. Four forked stakes, if more convenient, may be substituted for the logs and driven into the ground so as to receive the ends of the poles and stretch the sacking. The space in the bag between the poles should be filled with dry grass, leaves, evergreen boughs, or moss, which will give it the warmth and softness of a straw bed. By this arrangement you have an extempore bedstead, raising you above the cold, damp ground, and a bed as good as the best. For bed-clothes, the best contrivance is a bag made of wide, firm Canton flannel, six and a half feet long, open at one end. Let the tired hunter insert himself in this bag feet foremost, and he will need no "tucking up" to keep him comfortable even on the ground or in the snow; and if he is fortunate enough to be perched on such a bed as is above described, in a tent well buttoned up, with a friendly stove at his feet, the cry of the loon, the howl of the wolf, or the scream of the panther, will hardly disturb his slumbers.*

CAMP-CHEST.

A chest made of light materials, two feet nine inches in length, eighteen inches in width, and fourteen inches in depth — not larger than an ordinary trunk — will hold in transportation the stove with its pipe and all its furniture, the bed and bedding, the tent and all its rigging, and in fact nearly the whole outfit that has been described. The cover of the chest should be made of two thicknesses of boards, five eighths of an inch thick, with double hinges, so that the upper lid can be turned back separately, and form with the other lid a good table.

COOKING.

It will not be expected that the trapper's larder will be supplied with all the varieties and luxuries that can be found at the St. Nicholas, or at a Saratoga hotel. But it will always

---

* For a winter campaign, we would recommend the addition of a woolen blanket. — EDITORS.

be a satisfaction to know that flesh, fish, and fowl, are fresh from their native elements, and have not hung in the market two or three weeks before coming on the table.

The ways of cooking in camp are as various as in the kitchen at home. Fresh fish can be fried in butter, lard, or the fat of the deer; or they can be boiled, or broiled and buttered. Venison can be fried, or broiled in cutlets, or roasted before a camp-fire in joints, or stewed *a la fricassee*, or boiled into soup with potatoes. Squirrels, ducks, partridges, woodcock, quails, pigeons, prairie fowls, and any other game that comes to hand, can be fried, broiled, or boiled as well in the woods as in the best hotel.

The very best way of cooking fish and fowl ever devised is familiar to woodsmen, but unknown to city epicures. It is this: Take a large fish — say a trout of three or four pounds, fresh from its gambols in the cool stream — cut a small hole at the neck and abstract the intestines. Wash the inside clean, and season it with pepper and salt; or if convenient, fill it with stuffing made of bread-crumbs or crackers chopped up with meat. Make a fire outside the tent, and when it has burned down to embers, rake it open, put in the fish, and cover it with the coals and hot ashes. Within an hour take it from its bed, peel off the skin from the clean flesh, and you will have a trout with all its original juices and flavors preserved within it; a dish too good, as Izaak Walton would say, "for any but very honest men."

Grouse, ducks, and various other fowls can be cooked deliciously in a similar way. The intestines of the bird should be taken out by a small hole at the vent, and the inside washed and stuffed as before. Then wet the feathers thoroughly, and cover with hot embers. When the cooking is finished, peel off the burnt feathers and skin, and you will find underneath a lump of nice juicy flesh, which, when once tasted, will never be forgotten. The peculiar advantage of this method of roasting is that the covering of embers prevents the escape of the juices by evaporation.

Everybody knows how to cook potatoes and make tea and coffee, and anybody fit for a trapper must "know beans," and how to cook them. But bread! asks the novice; what are we to do for bread? Well, we have good, sifted Indian meal, and we will put some into a basin or pail, add a little salt, pour on scalding water, and mix to the consistency of stiff batter. After our venison or fish is cooked, we will put this batter into the hot fat that remains, a spoonful in a place, leveling it down smoothly, and turning it over till it is "done brown." Such a Johnnycake, served up with butter and sugar, would tempt a man to leave the best wheat bread that ever was made.

JERKED MEAT.

If you have the fortune to kill a deer or a moose in warm weather, and have an over-supply of meat that is likely to be tainted, you can preserve it by the following process: Cut all the flesh from the bones in thin strips, and place them, for convenience, on the inside of the hide. Add two or three quarts of salt for a moose, and a pint and a half for a deer, well worked in. Cover the whole with the sides and corners of the hide to keep out flies, and let it remain in this condition about two hours. Drive four forked stakes into the ground so as to form a square of about eight or ten feet, leaving the forks four feet high. Lay two poles across one way in these forks, and fill the whole space the other way with poles laid on the first two, about two inches apart. The strips of flesh should then be laid across the poles, and a small fire of clean hard wood should be started underneath, and kept up for twenty-four hours. This process will reduce the weight of the flesh more than half, bringing it to a condition like that of dried or smoked beef, in which it will keep any length of time. This is called *jerked venison*. It is good eating, and always commands a high price in market. An over-supply of fish can be treated in the same manner. They should be split open on the back and the backbone taken out.

PREPARATIONS AGAINST INSECTS.

In the warm months, chiefly from the first of June to the first of September, woodsmen are annoyed by myriads of flies, gnats, and mosquitoes. These can be driven out of a tent by smoke, and can be kept out by buttoning all tight. But the trapper should also provide himself with a protective against these pests. A good preparation for this purpose may be made by warming about three ounces of hog's lard, and adding to it half an ounce of the oil of pennyroyal. This ointment, applied once in an hour or less, to the parts exposed, will give entire protection.

Another preparation can be made by mixing equal parts of common tar with sweet oil, applying as before. This preparation is by some considered the best, because it also prevents tanning, and is easily washed off with soap, leaving the skin soft and white.

THE SHANTY.

The tent which I have recommended is probably best adapted to the irregular operations of amateur sportsmen, the volunteers and guerrillas of the trap. The old regulars and veterans of the service always have built, and probably will continue to build, rude huts, called shanties, at various points in the region of their operations. Shanties are of two kinds, temporary and permanent. The temporary shanty is made by driving two forked stakes into the ground, laying a ridge-pole across, leaning many other poles against this, and covering the skeleton thus formed with bark or split boards. The permanent shanty is made of logs, laid one above another in a square form, joined at the corners by means of notches, and roofed over with split logs formed into troughs, and placed in this form : ᘐᘐᘐᘐ. The crevices should be stopped with clay or moss. At one end a rude fire-place and chimney of stone should be built, the latter reaching just above the top of the shanty.

TRAPPING LINES.

Trapping, when carried on systematically and on the large scale, has, like an army, its lines of operation, its depots of provisions, and its arrangements for keeping open its communications with its base. The general proceedings of a regular trapping campaign may be described as follows: The trapping

The Home Shanty

company, which consists generally of two, three, or four persons, start out a little before the trapping season commences; select their lines, extending into the woods frequently from thirty to fifty miles; carry along, and deposit at intervals on the line, traps and provisions; and build shanties at convenient points, for sleeping-posts and shelters from storms. These preparations sometimes require several journeys and returns, and are made in advance of the trapping season, so that, when trapping commences, all hands may have nothing else to attend to. If the line extends directly from a settlement, so that it has what may be called a home-base, none but rude, temporary shanties are built; and once in about ten days, during the season, a man is sent back to the settlement, to carry out furs and bring back provisions. But, if the line commences so far from the frontier that such return-journeys are impracticable, then, besides the temporary shanties, a more substantial and permanent hut, called the home-shanty, is built at some point on the line, for depositing furs, provisions, and other valuables; and this becomes the base of operations for the season. A boy is sometimes taken along to "keep shanty," as trappers say, *i. e.*, to remain at the home-shanty as housekeeper and guard. Such a resident at the main depot is very necessary, as bears and other wild animals (not to mention fire and human thieves) have a habit of breaking into an unguarded shanty, and destroying everything within reach. Prudent trappers rarely leave furs in a shanty alone, even though it is strongly barricaded. If they cannot carry them out to the settlement, and have no boy to "keep shanty," they generally hide them in hollow trees. At the close of a season, if the party are satisfied with their line, and intend to trap on it another season, they hide their traps under rocks, where they will not be exposed to the fires that sweep the woods in dry times.

CONCLUSION.

The trapper's art, like that to which I have so often compared it — the art of war — is, or should be, progressive. It is evidently yet in its infancy, and has hardly begun to emerge from the narrowness and ignorance of mere individual cunning, into the liberal inventiveness and broad combinations which will come when trappers shall gather into conventions, compare experiences, and avail themselves of the help that all sciences are ready to give them. All that I can claim to have done in the preceding pages is, the presentation of the art of capturing animals, curing their skins, and living in the woods, as it now stands, or at least as I understand it.

Deer breaking cover.

THE TRAPPER'S FOOD.

By T. L. PITT.

THE trapper on his expeditions must often depend on his rifle or trap for subsistence. I will indicate the leading kinds of game which supply his wants, and methods of obtaining them.

DEER.

Among food animals, Mr. Newhouse has noticed the Deer and Moose. These are the trapper's most desirable game throughout all northern countries. In America, we have the common Red or Virginia Deer; the Black-tailed Deer, two varieties; the Long-tailed Deer of the Pacific slope; the Wapita or Stag, once distributed over a large portion of the Continent, but now found principally west of the Mississippi, in Oregon and Washington Territory, and in some parts of British America; the Moose; two varieties of the Caribou or Reindeer, in British America; and the Mule Deer of the Rocky Mountains. In Europe and Asia are the Moose or Elk; the Stag or Red Deer; the Fallow Deer; the Reindeer; the Persian or Indian Red Deer; the Thibetan Stag; the Sika of Japan; the Axis Deer of India; besides many other varieties in Asia, especially in the southern part.

The best method, and the one most to be relied on by the trapper, for hunting Deer, is what is called the "still hunt." The practice of hunting by boat and torch on lakes and streams, at night, is only adapted to the summer months, when trapping is out of the question, and when Deer should not be hunted, it being their breeding season. The plan of running Deer into lakes with dogs, though often practiced, is discarded and condemned by the best Deer hunters, as it tends to make the Deer wild, and to drive them into other regions. It may be resorted to when necessary, but cannot be recommended. It involves also the keeping of a dog which is generally of little use for any other purpose, and is a constant bill of expense. "Still hunting" is practised by finding the fresh track of the Deer, and carefully and noiselessly following up the trail till the location of the animal is discovered, when, by careful approach, a good shot can generally be obtained. Practiced Deer hunters become wonderfully keen, accurate, and successful in the still hunt. Messrs. Holland and Gunter, of Hastings County, Canada West, — the former of whom is one of the most accomplished deer-hunters in Canada, — give the following directions for this method: —

"For still hunting, the hunter should provide himself with a good rifle and a pair of deer-skin moccasins. When finding the trail he should walk carefully, and keep a good lookout ahead, as Deer are always watching back on their trail. When routed they almost always stop on hills. In order to get within gunshot it is necessary to circle round and come up toward them in front or at the side — always circling to the leeward side, as their sense of smell is very acute. The Deer, when the early snows come, usually get up and feed till about ten o'clock, A. M.; then they lie down till about three o'clock, P. M., when they start on a rambling excursion till near the next morning. In these excursions they almost always return to the place from whence they started, or near to it."

In still hunting, if buck, doe and fawns are found together, shoot the doe first, as in that case the buck will not leave the place till you have had opportunity for another shot. Deer when they lie down, turn off from their run-way, or track, and take a zigzag course back a short distance. They lie in a position which commands a view of the back track.

THE BUFFALO.

This animal is the great resource of the hunter for food on the western plains. Their range is from Texas to within about twenty miles of the Great Slave Lake. But few, however, reach this latter limit. They are seldom found west of the Rocky Mountains, and never, at the present time, east of the Mississippi. They are migratory animals, moving north in the spring with the advance of vegetation, and south in the autumn with the decline of pasturage. They move in large bodies, grazing as they go, and breed on the march. They usually reach the Platte River on their way north about the last of May. On their return they reach the same river in September. A few probably winter north of that latitude. These are mostly animals that wander from the great herds and get lost among the valleys in the mountains. On the uplands the Buffaloes live on a short, fine grass, called Buffalo grass. On the low lands they feed on a coarse, high grass. On their general march they move in a scattered, grazing order. Only when disturbed do they herd together and move in compact masses. When moving in the mass they stop for nothing, rushing through ravines, swimming rivers, and trampling all ordinary obstacles under foot. It is exceedingly dangerous to get in the way of a drove when on the rush. They should only be approached on the outskirts. Cows run the fastest. The bulls generally take the lead when the rush is made, but are soon outstripped by the cows. The cows and calves keep on the outskirts of the drove. A drove lie down where night overtakes them.

The common way of hunting the Buffalo is on horseback, as a person on foot cannot approach them without screening himself. Experienced hunters prefer a largest sized or eight inch navy revolver for hunting them. A breech-loading carbine or rifle, is also a good weapon. Find a drove feeding. Approach them from the leeward side, otherwise the animals will scent you and move off. They are not disturbed by a horse as long as they do not scent the rider. Lie down on the horse and let him gradually work his way into the drove. Select a cow and approach her on the left side if you have a pistol, on the right side if you have a rifle, in order, in either case, that you may have the best opportunity for using your weapon. Shoot for the heart, which lies, comparatively, very low. The ball should be aimed just back of the fore leg, a few inches above the brisket. The ball if aimed right will generally go through, and the animal will soon bleed to death. New hunters are liable to aim too high, being deceived by the height of the hump on the shoulders. They suppose the heart is near the middle of the space from the top of the shoulders to the brisket; it is some distance below that point. The danger in Buffalo hunting for beginners, is in getting too far into the drove. As soon as an animal is wounded the rest take the alarm and close round, and if the hunter has not secured a way of escape he will probably be run down and both horse and rider destroyed. When chasing a Buffalo and shooting on the gallop, the hunter should bring his horse into time with the animal; otherwise he will probably miss his aim. He should fire just as the horse and the Buffalo strike the ground with the fore feet.

The cows are best for eating. The tongue and tender-loin are preferred, the rest of the meat being rather coarse, especially that of the bulls, unless the animals are fat. It is, however, all eatable, and somewhat resembles beef, but has a strong, peculiar, wild flavor of its own. Much of its reputation may be due to the good appetites of those who hunt it.

The cows furnish the Buffalo robes of commerce, the skins of the bulls having no fur on the hinder parts, and only the long coarse mane in front; their hinder parts are covered with short hair. The bull-skins make a coarse kind of leather, used by the Indians of the plains to cover their wigwams and for other purposes.

THE MOUFFLONS OR GREAT HORNED SHEEP.

There are several species of wild sheep which are of some interest to the trapper. The first of these is

THE ROCKY MOUNTAIN SHEEP OR BIG-HORN.

This animal is larger than the common sheep, being sometimes six feet long, about three feet high at the shoulders, and weighing nearly three hundred and fifty pounds. They are found throughout the whole range of the Rocky Mountains, from the 30th to the 68th degree of north latitude. The horns of the males are enormous, measuring over two feet ten inches in length round the curve, and being very large at the base. Their color is a rufous gray, except the rump, belly, and the inside of the hind legs, all of which are a grayish white. In winter they become lighter-colored. The hair is coarse and slightly crimped. Underneath the hair is a soft fur or wool. The Big-horn is, or becomes after contact with hunters, an exceedingly shy, wild animal. In the retired parts of the mountains where they have never been hunted, they are sometimes found quite tame and unsuspecting. They are gregarious and live in small flocks among the peaks and most inaccessible regions of the mountains, never descending into the plains. They subsist on mountain grass and herbage, and inhabit the rocky recesses. The young rams and the females herd together during the winter and spring, while the old rams separate in flocks, except at the rutting season in December. The rams fight fiercely with each other like common rams. The ewes bring forth one or two young in June and July.

The flesh of the Big-horn is excellent, superior to the best venison or the finest mutton. They can only be hunted successfully by the exercise of extreme caution and strategy in approaching them; and if only wounded by the first fire they retire to their recesses among the rocks, and there die, inaccessible to the hunter. Dogs are worse than useless in hunting them.

Another Moufflon is

THE ARGALI.

The Argali of Siberia and Central Asia greatly resembles the American big-horn, and some naturalists have regarded them as the same species. They are very large, being about four feet high at the shoulders and proportionately large in build. The horns of a full grown male are nearly four feet in length, measured along the curve, and about nineteen inches in circumference at the base. They rise from the forehead a short distance, then curve downward below the chin, then recurve upward and terminate in a point. They are mountain-loving animals and are found in the highlands and mountain ranges of Siberia and Central Asia. They are very fleet and sure of foot, and when disturbed rush to the most inaccessible places among the rocks and peaks. They are gregarious and live in small flocks. In the winter these flocks are sometimes enveloped in the deep snow-drifts. In such cases they lie quietly under the snow and respire through a small breathing-hole. The hunters eagerly hunt for these imprisoned Argalis, and spear them through the snow. At other times they are hunted with the same cautious strategy that is required in the case of the big-horn.

THE PRONG-HORN ANTELOPE.

This animal abounds on the western plains of the United States. It is the only species of Antelope in North America. It is of nearly the same size as the Virginia deer. They differ from all other Antelopes in having a prong or branch on each horn. This prong is situated about the middle of the horn on the anterior face. The tops of the horns curve inward and backward, forming a small hook like those of the chamois. The legs of the Prong-horn are long and slender, the ears long, narrow, and pointed, and the tail short and bushy. The whole form is stately, elegant, and graceful. The color of the upper parts is a yellowish-brown; the under parts, with a patch on the rump, are grayish-white. Their favorite haunts are the low prairies adjoining the covered woody bottoms. They are also found on the upland prairies, and along the rivers and streams. They swim well. They sometimes congregate in large flocks; at other times only one or two are seen. In the winter the Indians take advantage of their congregating together and hunt them by a "surround." The manner of doing this is as follows: A large number of Indians distribute themselves around the Antelope at such a distance as not to alarm them. Then they advance with cries and noise from all sides. The Antelope, instead of endeavoring to escape, herd closer together in their fright, and suffer themselves to be beaten down with clubs. In this way great numbers are sometimes killed. Though very wild and shy, the Antelope is full of curiosity. Any novel object at-

RUFFED GROUSE.

tracts their attention. At length curiosity overcomes timidity, and they advance to examine it. The hunter takes advantage of this trait. Concealing himself, he attaches a red or white flag to his ramrod, and with it attracts the animal within range of his rifle. Their sense of smell is very acute, consequently the hunter should always keep to the leeward of them. They are among the fleetest of all animals. They inhabit all the western part of North America from the Saskatchewan to the plains of New Mexico. Their flesh is inferior to that of the deer.

SQUIRREL HUNTING.

Squirrels are usually considered "small game" by trappers, requiring more ammunition to kill them than they are worth. There are times, however, when they furnish an acceptable addition to woodland fare. The best way to hunt them is this: Find a piece of woods where they abound. Go into the woods and seat yourself on a fallen tree or rock. Remain motionless and quiet. Soon you will begin to hear the Squirrels at their work or see them among the trees. By patience and the most quiet strategic movements you will soon get a shot. Several may sometimes be shot from one position, in a short time. The great point in Squirrel hunting is to avoid all unnecessary moving about.

GROUSE.

The Grouse family furnishes the trapper his most desirable winged game, throughout the world. In this country the leading kinds of Grouse are the following:—

THE RUFFED GROUSE.

This bird is known in New England as the Partridge, and in some of the Southern and Middle States as the Pheasant. Neither of these names is the proper one, for this bird belongs to neither the partridge nor the pheasant families. The wild turkeys are the only representatives of the pheasant family in North America; and the so-called quail is our true partridge. Let us hereafter, not only as naturalists, but as hunters and trappers, call this noble bird by its true American name

— Ruffed Grouse. There are three species of the Ruffed Grouse: the common species which inhabits the country from the Southern States to Labrador and the Saskatchewan; the Oregon or Sabine's Grouse of the Rocky Mountains and the Pacific slope, and the Allied Grouse inhabiting the Rocky Mountains northward to the frozen regions. The Oregon Grouse is much darker and redder than the common species. The Allied Grouse is of a light gray color, and is smaller than either of the others. All are excellent for the table. Ruffed Grouse are generally found in small flocks, except where they have been much hunted. In the latter case more than two are rarely found together. They delight in upland and mountain forests, where springs and small brooks abound. They are particularly fond of the high, sloping banks which border on such streams. These are their favorite feeding-grounds. Their flesh is white and unsurpassed in flavor by other Grouse. They should be hunted with a trained dog. Sportsmen prefer cockers. In the back woods they may occasionally be hunted with moderate success without a dog; but such hunting is generally tedious and uncertain. They are easily snared by building a low fence of twigs with occasional openings, large enough to permit a bird to pass through, and placing a slip-noose across the opening. The noose should be made of small copper wire. Some hunters prefer to attach the noose to a spring-pole.

THE PINNATED GROUSE.

This species is commonly known as the Prairie Hen. They formerly existed in great numbers in the Atlantic States, but are now mostly confined to the prairies and plains of the West, east of the Rocky Mountains, within the limits of the United States. They differ from the ruffed grouse in preferring the open country to the forests. They choose the dry lands for their habitat, avoiding as far as possible marshy or wet places. They depend for their drink on the dew which they collect from the leaves of plants. In color the Prairie Hen resembles the ruffed grouse, but its markings are different. It is about nineteen inches long and when in good order, weighs about three and a half pounds. Its meat is dark-colored but fine flavored. The neck is furnished with a pair of supplemental wings, about three inches long; underneath these are orange-colored air-sacs, which can be inflated to the size of a medium sized orange. Audubon says that when these sacs are "perfectly inflated, the bird lowers its head to the ground, opens its bill, and sends forth, as it were, the air contained in these bladders in distinctly separated notes, rolling one after another from loud to low, and producing a sound like that of a large muffled drum. This done, the bird immediately erects itself, refills its receptacles by inhalation, and again proceeds with its 'tootings.'" These tootings can be sometimes heard at the distance of a mile. Their food consists of the seeds of the sumach, grapes, grain, wild strawberries, cranberries, partridgeberries, whortleberries, blackberries and young buds. They also eat worms, grasshoppers and insects, and in winter feed on acorns, the tender buds of the pine, clover leaves, and, when possible, frequent grain stubbles. They are best hunted with a trained dog.

THE SHARP-TAIL GROUSE.

This bird is allied to and greatly resembles the preceding. It takes the place of the prairie hen in the far West, on the plains that skirt the eastern base of the Rocky Mountains. It avoids the highlands and mountains, and has its habitat on the prairie and open grounds. There they congregate in flocks, feeding on wild rye. Near settlements they frequent grain stubbles. They hybridize with the pinnated grouse, and are equally excellent for food. They are destitute of the gular sacs on the neck. Their range extends northward into British America. In the far North there is another species called the Arctic Sharp-tailed Grouse. They are about the same in size as the preceding, but are darker in color, being black where the other is brown.

THE COCK OF THE PLAINS.

This is the largest of the American grouse. Its common name is the Sage Cock. Its habitat is chiefly on the western plains on both sides of the Rocky Mountains where the wild sage or artemisia grows. It feeds on the leaves of this plant, which being very bitter, give the flesh an unpalatable flavor. In the autumn, however, they frequent the streams of the Columbia, where they feed on the pulpy-leaved thorn. At this season their flesh is good. The males have large, orange-colored gular sacs on the sides of the neck, which they inflate, and by expelling the air produce " a loud, grating noise, resembling hurr-hurr-r-r-hoo, ending in a deep hollow tone, not unlike the sound caused by blowing into a large reed." Their general color is light brown, marked with black, dark brown and yellowish white. They are large, weighing frequently five or six pounds. The tail is long and pheasant-shaped.

THE DUSKY GROUSE.

The species next in size to the preceding is the Dusky Grouse, sometimes called the Pine Grouse. It is an inhabitant of the Pacific slope and of the Rocky Mountains from the Columbia River to Texas. They are supposed to be partially migratory, leaving their accustomed haunts in November and being absent until spring. Their flesh is said to be excellent, having a slight pine flavor, which is not disagreeable. The Dusky Grouse is easily captured. Their habit is to spend most of their time on the ground. They lie close till almost stepped on, and when disturbed take refuge in the nearest tree, alighting among the branches and remaining motionless. Richardson's Grouse resembles the Dusky Grouse, but its habitat is in the Rocky Mountains from the South Pass northward.

THE SPRUCE OR CANADA GROUSE.

This bird is found from the northern United States to the Arctic Sea, and from the Atlantic nearly to the Rocky Mountains. Their favorite habitat is the thick evergreen swamps. They are less wild and shy than the other kinds of Grouse, and are said to be easily tamed. When confined, they feed readily on oats, wheat, or other grain. Their flesh is quite dark, and in winter, when they feed on the leaves of evergreens, is unpalatable. In the season of berries it is much better flavored. In the Rocky Mountains a species of Grouse is found which closely resembles the Spruce Grouse, except that its habitat is in the mountains rather than in the swamps. This species is called Franklin's Grouse.

PTARMIGAN.

Allied to the grouse, and known by the name of Snow Grouse are the Ptarmigans. They inhabit the northern parts of both continents, especially the cold snowy regions near or within the Arctic Circle. They differ from the common Grouse in having their legs and feet completely feathered, leaving no portion of the body exposed except the bill and nails. They all turn white in winter, but in summer are beautifully mottled with various colors. Only one species has its habitat within the limits of the United States. This is the

WHITE-TAILED PTARMIGAN

of the Rocky Mountains. They inhabit the regions of eternal snow, and only descend to the lower levels to breed. Not much is known of this species except that they are wild and shy. Their color in winter is the same as their snowy surroundings, and in summer resembles that of the moss and lichen covered rocks.

THE WILLOW PTARMIGAN.

This is an important bird and furnishes a large amount of food to the inhabitants of British America, particularly to the natives and trappers of the Hudson's Bay territory. In winter they sometimes enter the limits of the Northern States, and their range is from Nova Scotia and Newfoundland to the Rocky Mountains and the Arctic Sea. They breed plentifully in Newfoundland, Labrador and the fur countries. They live mostly on the ground. They are wonderfully prolific, and vast numbers of them are found and captured in some localities. Hearne, who travelled and explored in the Hudson's Bay region nearly a hundred years ago gives the following account of them: —

"They are by far the most numerous of the grouse species that are found in Hudson's Bay, and in some places, when permitted to remain undisturbed for a considerable time, their number is frequently so great as almost to exceed credibility. I shall by no means exceed the truth if I assert that I have seen upwards of four hundred in one flock near Churchill River; but the greatest number I ever saw was on the north side of Port Nelson River, when returning with a packet in March. At that time I saw thousands flying to the north, and the whole surface of the snow seemed to be in motion by those that were feeding on the tops of the short willows. . . . In summer they eat berries and small herbage. Their food in winter being dry and hard, makes it necessary for them to swallow a considerable quantity of gravel to promote digestion, but the great depth of snow renders it very scarce during that season. The Indians, having considered this point, invented the method now in vogue among the English of catching them in nets by means of that simple allurement, a heap of gravel. The nets for this purpose are from eight to twelve feet square, and are stretched on a frame of wood, and are usually set on the ice of rivers, creeks, ponds, and lakes, about one hundred yards from the willows, but in some situations not half that distance. Under the centre of the net a heap of snow is thrown up to the size of one or two bushels, and, when well packed, is covered with gravel. To set the nets when thus prepared requires no other trouble than lifting up one side of the frame and supporting it with two small props about four feet long: a line is fastened to these props, the other end being conveyed to the neighboring willows, so that a man can always get at it without being seen by the birds under the net. When everything is thus prepared, the hunters go to the adjacent willows and woods, and, when they start the game, endeavor to drive it into the net, which at times is no hard task, as they frequently run before them like chickens; and sometimes require no driving, for, as soon as they see the black heap of gravel on the snow they fly straight toward it. The hunter then goes to the end of the line, and when he sees that there are as many about the gravel as the net can cover, or as many as are likely to go under at that time, with a sudden pull he hauls down the stakes and the net falls on the snow, and incloses the greater portion of the birds that are under it. By this simple contrivance I have known upwards of three hundred caught in one morning by three persons."

The weight of the Willow Ptarmigan is about one and a half pounds. Another species called the Rock Ptarmigan is

Canada Goose.

found in British America, throughout nearly the same range. They are smaller than the Willow Ptarmigan, and congregate together in great numbers in the open grounds in winter.

EUROPEAN GROUSE.

The principal Grouse of Europe are the Capercaille and the Black Grouse. The former is a very large bird, about three feet long, and inhabits the wooded portions of Northern Europe, especially those of Sweden and Norway. The Black Grouse is also quite large. They are abundant in Sweden and Norway, and Northern Europe. Several species of ptarmigan are also found on the Eastern Continent.

WATER FOWL.

Two families of water fowl are of considerable importance to the trapper. These are the Ducks and the Geese. The former are so familiar as to need no description. I will merely enumerate the species that are esteemed for the table. These are divided into the sub-families of Sea Ducks and River Ducks. The latter principally frequent the inland waters, and are all good for the table. They are the Pin-tail, the Mallard, the Black or Dusky, the Shoveller, the Muscovy, the Wood, the Widgeon, the Green-winged and Blue-winged Teal, and the Gadwall. Of the Sea Ducks, only the Redhead, the Canvas-back, and the Ring-neck are much esteemed. The two first are excellent. All the vegetable-eating ducks are adapted to table use; the fish-eaters are poor. They are hunted with decoys in the early spring and fall, and in summer with punt boats. A heavy shoulder gun with wide bore and long range is used.

The Wild Geese spread over the whole of this Continent and abound in Europe and Asia. They breed in the far north. They migrate north in the early spring, and return south late in the fall. In the fur-countries of British America they constitute the principal summer food of the inhabitants, and are salted down in great numbers for winter use. They are shot from behind screens on the margins of lakes and rivers. The hunters decoy them within range by imitating their cries. Tame Geese may also be used as decoys.

FISHING IN AUTUMN AND WINTER.

By T. L. PITT.

In the fall, beside the ordinary methods of fishing with hook and line, which are too familiar to need description, the trapper may have opportunity for spearing salmon-trout on their spawning beds. This operation, to be successful, requires these preliminaries: 1, plenty of fish, and good spawning beds; 2, a good canoe or boat; 3, a good spear; 4, a good jack; 5, plenty of fat pine and white-birch bark; 6, favorable weather; 7, an expert spearman. The jack is a sort of concave gridiron structure, made of wire or iron rods, and placed on an upright post about three feet and a half high, in the bow of the boat. In the jack the fat pine and birch bark are burned to give light to the spearman and those who paddle the boat. Fat pine, is pine that is full of pitch, and is usually found in the knots and roots of fallen and decayed trees. The spear should be made with five barbed prongs, about five inches in length and three fourths of an inch apart, and set on a line with each other. The prongs should be made of the best steel, well tempered. The four outer prongs should be barbed on their inside edges. The middle prong on both edges.

The practical operation of spearing is as follows. Having arrived on the spawning ground the spearman kindles the fire in his jack, as soon as it is dark enough. He then stands near the bow with spear in hand, and peers keenly down into the water for the desired fish. The paddler stands near the stern, and follows the directions of the spearman in paddling and guiding the boat. The spearman must stand firmly in the boat, and in striking must allow for the refraction of the light in the water. If a fish appears to be one foot below the surface, he is really much deeper, and if you strike at the apparent depth you will surely miss him. The spearman, however, soon learns by his mistakes to make the proper allowances; and when he learns this, and attains self-poise, calmness, and quickness of movement, will be successful.

The places which salmon-trout choose for spawning are on the westerly-looking shores of lakes, and the coasts of islands, where the slope is gentle, and covered with large, clean gravel and rocks.

FISHING THROUGH THE ICE.

In winter, brook trout may be caught on sand-bars, where the water is two or three feet deep, and lake trout in deeper water, by cutting holes through the ice and fishing with hook and line. One person may fish with several lines in different holes, by using *tip-ups*. These contrivances are made in this way: Take a strip of shingle, or board, two inches wide and twenty inches long. Bore a hole through it near one end. Through the hole insert a stick, long enough to reach across the hole in the ice. Then fasten your fish-line to the short end of the strip, and drop the hooks into the water. When a fish bites the long end will tip up and attract your attention. Bait with any sort of meat. Cow's udder makes excellent bait on account of its toughness. Salmon-trout are caught in the same way, only in deep water, where the banks are bold.

NET-FISHING IN WINTER.

Fishing may be performed under the ice with gill-nets in the following manner: The net is fastened with loops or rings to a long, smooth pole. The loops or rings should be large enough to slip easily along the pole; or if preferred the net may be hung on a rope, each end of which is fastened to the ends of the pole. Two holes are then cut in the ice, the length of the net apart, and the pole and net are sunk under the ice and fastened between the holes. Two cords should be attached to one end of the net, near the pole, and brought up and fastened above the ice, one through each hole. When the net is to be examined, it is drawn together at one end of the pole, by means of one of the cords, and taken up through the hole, which should be kept open. After removing the fish, the net is dropped back into the water and spread out along the pole again by means of the other cord. Some fishers prefer to swing their nets on a rope without any pole. In this case holes should be cut through the ice, six or eight feet apart, along the line of the net, and the rope brought up and passed over sticks laid across the holes. The net should also be arranged with cords, so that it can be examined through the middle hole, by drawing it from each end of the rope to that point.

NOTES ON TRAPPING AND WOOD-CRAFT.

By "F. R."

[THE following article was written by a practical trapper, in response to an invitation from Mr. Newhouse, and partly as a criticism on our first edition of the "Guide." As his suggestions are the result of actual experience, they will be found interesting to the trapper, whether strictly followed or not. — EDITORS.]

IT would be a great advantage to young and inexperienced trappers if they could have descriptions and engravings of the foot-prints or tracks of animals. Even those skilled in the trapper's art are at times deceived and led off on some "wild-goose chase" for want of such information. As an instance, I will relate the following story: Once, when a boy, hunting in a well-settled region in the State of New York, I discerned otter signs. The otters appeared to have no regular abiding-place, but wandered at will, up and down the stream, a distance of some four or five miles, between two lakes. There were five or six of them, and so "shy" and wary were they, that they defied all attempts to trap them. Having at length discovered that they lurked near a certain "deep hole" in the creek, early dawn found me near the spot, with my gun well charged with buckshot, and accompanied by my two dogs, with whose assistance I expected to get the otter out of the water, *when I killed him*. There was a piece of swamp which I had to cross, in order to reach my post of observation. This swamp lay so open to the creek that I crawled across it on my knees, to escape, if possible, the notice of the otter, should any be lurking near, dragging myself along through the deep and fresh fallen snow, each leg as it trailed making a deep gouge, and both forming two long, parallel gutters. In each of these gutters walked a dog, soberly enough, much obliged to me, no doubt, for thus making him a path. I reached my post, and spent the morning without observing any thing unusual. Toward noon I arose and was about to start for home to dinner, when I descried two men making their way toward me across the marsh, evidently much excited, eagerly gesticulating and inciting one another to haste. Seeing me they stopped, and asked me whether I had "seen the otters." Upon my replying in the negative, they laughed incontinently, declaring that I was blinder than a bat; that I must have been asleep, &c. "Why," said one, observing my astonishment at their conduct, "here are their tracks, covering yours, scarce a rod from where you sit. See! here they've taken to water. We first came upon their trail as we were crossing the swamp there. By their tracks, I make them to be two of the biggest critters I ever so much as hearn tell of. We hurried on, thinking we might perhaps catch them ashore."

After some further conversation, they hurried on down stream, leaving me, to use a common phrase, "rather mixed." I was certain that no otter had come within many a rod of me. I had watched eagerly for a single wave or ripple in the placid waters of the stream from under the snow-covered bushes, whose pendent boughs almost reached the water and formed a curtain to the opposite bank. There was no sign, nor had there been — not a trace. I was quite sure I could not have passed an otter trail without noticing it — the unmistakable scoop of his long, stovepipe-like body, with paw marks interspersed along it. I retraced my steps to the spot where I first struck the creek, after crossing the swamp, which was the spot where they had said the otters had taken to water again. Truly, there was their trail, a couple of them, big ones at that. I called the dogs, and showed them the tracks. To my surprise they were nowise excited about it; "sniffed" and turned away. Extraordinary conduct! — which raised a latent suspicion. I doubted — thought — then light flashed upon me, and I burst into a hearty laugh. It was a great joke. Of course you understand it all. The long gouges which my knees had made in the yielding snow they had mistaken for the drag of the otters' bodies; the prints of the dogs' feet for the otters' paws. You may say they were superficial observers. Excitement will have its effects, and nothing but correct information can in such a case counteract it. "Knowledge is power."

The print of a raccoon's paw greatly resembles that of the bare foot of a young child. It is easily recognized. The bear, woodchuck, and skunk are also plantigrade; but the print of their paws has little resemblance to the "coon's."

Otter will not eat bait, as a general thing; but they will smell of it, which is frequently just as good. Some stale meat, or better, fish, will attract them, especially if it is placed in a queer, unusual position, hung from a bush or stake, so as to attract their attention. Inquisitive as they are, the trapper should take care that the object or bait excites their curiosity without alarming them.

I have been informed by experienced trappers that a wolf-trap should be well rubbed with the green leaves of the male fern or "brake" when they are to be had. They give a humid, earthy smell to the trap, and the juice, when it evaporates, appears to carry off all scent of human contact. I suggest, however, that if trappers would lay out a little more money in buckskin gloves they would be well remunerated. The contact of the bare hand with the trap is very objectionable; you might as well hold out a noose and call a wild horse to put his head in it. The gloves should only be used when handling the trap. Some rub the traps with blood, when trapping carnivorous animals; others substitute herbs, as skunk-cabbage, &c., for all animals. For the bear, the Indians say, the best bait is skunk-cabbage. They are said to be very fond of it. I cannot verify this, for I have never had an opportunity to try it. It would take as sturdy a pine-bender as him that Theseus slew, to make a spring-pole that would raise a bear beyond wolf reach — for wolves will attack and devour even a bear, wounded and hampered.

The raccoon may frequently be taken during a hard frost, by cutting a hole in the ice on any stream which may be near his habitation. A trap set in this, will be almost sure of him. He will rise at midnight to paddle in the water, though the temperature stands at zero. Hence his Latin generic name of "Lotor."

I think that a live chicken is the best bait possible for the wild cat, and also for all feline animals. Fresh, *bloody* meat, however, of any description, is very enticing.

Till lately I have strongly adhered to the opinion that a "Black Lynx" was "dyed in the wool" — *after death.* Recent researches have almost made me doubt. I have received assurances from men whom I think reliable, that there is, or has lately been, such an animal in existence. How it could have escaped the sharp eyes of our naturalists, I cannot imagine. It is represented as being of large size, almost as large as the black bear; in form and general habits resembling the ordinary Canada lynx; but is said to be as ferocious as the Canada lynx is timid. The hair is said to be thick, long and shaggy, and as black as Erebus. It is also said to have great local attachments, never leaving the impenetrable wilderness of swamp which it inhabits. The Indians have many wild and curious legends or traditions which perhaps refer to this animal. He is doubtless — if he exists — the "*Lunxus*" or devil of the Indians of Maine. The "Black Lynx" is said to be able to throw a full grown sheep across his shoulders and make off with ease. "All the beasts of the wilderness dread him, and man himself cares not to invade the retired fastnesses of the gloomy forests where he rules absolute monarch."

Our backwoodsmen are almost as remarkable for their "yarns" as Jack Tar, and they are generally about as reliable. Did you ever see the pelt of a "Black Lynx," or of any other similar dark-colored animal? It must be a myth.*

The offensive smell of skunk, may be removed from clothes

---

* Your "Black Lynx" is probably the wolverene, modified and exaggerated by the imaginations of the trappers or hunters who caught a glimpse of it. The wolverene is the Indian Devil, and is so called by the Indians of British America. It is a very troublesome, sagacious, and destructive animal to the trappers, in the wilds where it dwells, but most of the extraordinary stories told of it are probably "yarns" like those formerly related by trappers of the beaver. — EDITORS.

by wrapping them in fresh hemlock boughs; in twenty-four hours they will be cleansed. They should be left out at night. I have known many who preferred the smell of the skunk to that of the musquash. As to eating a skunk, if other game is not to be had, I should not be fastidious. A skunk properly dressed and cooked is good eating.

Some think the flesh of the woodchuck or "groundhog" excellent, especially in the fall. He should be carefully skinned and cleaned immediately after death. Some dark strips of granular, brown fat, which lie along the inside of the animal's legs, should be carefully cut away, or the flesh will be spoilt. I have at times found the woodchuck up a tree, almost always in iron-wood trees. It is hard to dislodge them; they hold on like grim death, and cannot be shaken loose. What induces them to climb I cannot tell; they never appear to have any thing to do there. They get up amongst the small branches, and much resemble a knot or "bunch" of the wood. Their color also corresponds well with the bark of the iron-wood, and renders it difficult to detect them. I have been informed that they will climb hollow trees at times to escape pursuit, and that it is almost impossible to dislodge them by manual force. The rabbit, also — an animal which from its peculiar conformation would not be suspected of climbing — has frequently been found in the hollows of trees. It is supposed to climb like the old chimney-sweeps, being found with its back braced against the side of the hollow. By rabbit, I mean the small brown hare peculiar to this country. Their habits are similar to those of the great white or northern hare. They will sometimes inhabit a deserted woodchuck hole.

For deer and moose — though I do not believe in trapping these animals except for food — I consider the brush fence, noose and spring-pole the best method of catching them. A rope is the most simple and portable trap, and it is always useful. The Indians have a method of calling the moose with a horn of birch bark, producing a sound resembling the lowing of the cow, alluring the bull to destruction.

As to "life in the woods" the old Cromwellian motto,

"Trust in God, and keep your powder dry," is most excellent. I advise those who are wise enough to wish to follow it, to use the flat tin powder cans, with metallic caps screwing down water-tight. The Hazzard and Dupont powder comes in such cans — pounds and half pounds. I have found that three dr. of Dupont's No. 2 (coarse ducking powder) is equal to four drs. of Hazzard's ordinary grain in strength. I use a twelve gauge duck gun. I think No. 4 shot is a good size for such game as ducks. With Ely's S.S.G., green cartridge, or large buck-shot and a twelve gauge gun, you can generally get all the deer you want. I consider No. 6 the best size shot for full-grown grouse. No. 8 does very well for smaller birds, woodcocks, &c., and red squirrels. I consider four (4) dr. of Hazzard's powder, and from one and one quarter ($1\frac{1}{4}$) to one and one half ($1\frac{1}{2}$) ounces of shot the proper load for a twelve gauge gun. At least it is for mine.* An iron ramrod should not be used; it wears the muzzle of the piece, and makes it scatter. Brass might do, if a metallic rod is considered a desideratum. Being softer than iron the wear would chiefly fall upon the rod. Hunters cannot be too careful to keep their salt away from their powder; it absorbs moisture and imparts it to the saltpetre of the powder. Here I will qualify my praise of water-proof tin cans for powder. They are the best things that can ordinarily be had for that purpose. But I would not advise any one to hide or *cache* powder in such a can. A week, aye, a few days, might suffice to turn your powder into a black, unctuous mud. The metal appears to attract moisture, and though the can may be impervious to any sudden shower or drenching, by some means, if long exposed, the moisture will get in. I think that a *horn*, plugged with pine wood, which has been boiled in a mixture of rosin, wax, and tallow, and the joints varnished, will be quite water-proof. I have known a horn of powder lost in the woods, and exposed for weeks (wet weather having intervened), to be dry and uninjured. A copper flask is worse than a tin can, in this respect. I prefer a horn flask,

---

* For large animals the charge of powder may be increased from one half, to one dram.

with a patent water-proof safety top and German silver mountings; but they are scarce and costly. The lightest and best camp-kettle is of "pressed tin." One of from three to four quarts is worth about one dollar, and is sufficient for two or three persons. It is very light and convenient, and should have a lid or cover with a wire handle which will fold down sideways, so that when inverted it could be used as a dish. The rim of this lid, or dish, should be quite broad, so as to make it capacious. It might be used to hold a portion of the contents of the kettle, mush or potatoes, &c. There should be a light wire chain attached to the handle of the pail by which to suspend it. For a hunting-knife, I use a bowie, and have found it an excellent tool. The sheath which comes with a knife is not good for much. I generally replace it with a strong wooden one, covered with leather. I take a flat piece of strong wood of the requisite shape, and saw into it lengthwise — the blade of the knife to be laid, edge first or down, into the space cut by the saw, and the back being towards the opening. This wooden case prevents the knife from cutting you, in case you should fall upon it, of which there is great danger where the ordinary pasteboard, leather-covered sheath is used. The sheath and knife should be attached to the belt by a frog, which should not be a permanent portion of the sheath. The army "camp knife" is a very nice thing for hunters; you have your spoon, fork, and knife in very compact shape — cost, one dollar and a half. A saw and an auger, with some large spikes, wrought nails, butts or hinges, staples, and a padlock or bolt are needed around the "home shanty." They tend to "make things comfortable" and safe. Your matches should be of the best; lucifers, or "Vienne water-proof." Their tips only are water-proof. I render them absolutely water-proof by dipping them in a solution of shell-lac in alcohol. This makes the "sticks" of the matches quite impervious to moisture. The solution of shell-lac, should not be too thick, or they will not burn well. When properly prepared in this manner, they may be immersed in water for twenty-four hours, and will then (if taken out and wiped dry) instantly ignite and burn well. As a final precaution, when they are so dry that there is no danger of their adhering to one another, I put them in a warm, dry bottle, with waxed or water-proof stopper or cork. This is the true way to carry any sort of matches.

I always prefer to put up matches, caps, &c., in several different packages or places, so that in case of accident all is not lost. This system should not, however, be carried to an extreme, as it is then both confusing and troublesome. Every thing should be plainly labeled. Boxes, &c., containing a miscellaneous assortment of stuff, should have a list on the outside, or on the inside of the cover.

As to provisions, I should leave out beans, which to be good require *time* for preparation, and instead, should carry a package of "*self-raising flour*" — wheat — an excellent article. With it you can make biscuit or bread on short notice. It is to be had of grocers generally, I believe, put up in six pound packages. Pork or lard, butter and sugar, are all the luxuries needed, except perhaps coffee and tea. You can fatten on them. Beef, butter, sugar, Indian meal, &c., are said to contain a great proportion of strength-giving food.

I quite agree with you on the subject of clothes, but will make a few suggestions. I prefer to have my boots first sewed in the ordinary manner, and then to have a light "Napoleon tap," pegged on with steel or copper nails. I soak a hot mixture of mutton-tallow, bees-wax, and rosin into the soles of boots, till they will absorb no more; such boots wear out slowly and the soles never get soaked or water-logged. The preparation I recommend is far superior to coal or common tar for this purpose; the boots do not "squeak" as those tarred will. There should be more tallow than wax, and more wax than rosin.

The trapper should always be provided with scissors, needles, pins, thread, &c.

Pork, bread, meal, &c., should be put up in neat boxes or bags, as nearly water and air-tight as possible, each neatly and legibly labeled, so as to pack easily and be known at sight, without rummaging. Bags should be painted or otherwise water-proofed. If lead paints are used, the article inclosed should be put in a paper bag *first;* white lead is, as all should know, very poisonous. Boiled linseed oil is apt to rot the material of linen or cotton bags.

As to cooking, I would advise all those who are at all fastidious as to their food to carry some vinegar and curry-powder, &c. I can assure you curry-powder improves a schytepoke wonderfully. Without further reference to *this* subject, I must say that onions come very good at times. Potatoes also are good, either baked or boiled; they are also healthy, portable, and convenient.

I can tell you of one of the nicest things known, namely, pork fritters; melt some lard in a saucepan or spider, make a stiff batter, but not *too* stiff either, of wheat or rye (boiled Indian meal might do); cut slices of pork, dip in the batter, and when the melted fat in the pan is quite hot, drop in your fritters. Cook till light brown. They are delicious. Try them any day; it is not at all necessary to have an appetite.* If some other drink besides water, tea, or coffee is considered absolutely necessary, carry lemons or oranges. With these, and plenty of sugar, joined with the cold clear water of some mountain spring, he who is not satisfied deserves never to be. Sugar and lemon-juice will make even warm swamp-water palatable to a thirsty man.

You give directions for the preservation of an overplus of venison, &c. This reminds me to ask how would you preserve a moose from wolves and other depredators in case you should be obliged to leave the carcass, to find help to remove it? I have heard it said, that the half-blown bladder of the animal suspended from the branch of a tree or bush over the carcass would answer; others say that a rope or even a cord loosely hung on the surrounding twigs would be sufficient, the wolf supposing it a trap.†

---

* We think a substitute for pork should be invented or adopted. It is about as bad for corrupting the blood as the alcoholic stimulants which the above writer condemns. Butter is good, but for all frying operations is less economical, and less satisfactory than olive oil. Pure, sweet olive oil, put up in air-tight or closely corked cans or flasks, would be portable and an excellent portion of the trapper's outfit. — EDITORS.

† Wolves will not meddle with a dead deer if it is laid by a log and a few

As for preparations against insects — they are of a very doubtful benefit. Those who wish to be comfortable, had better leave *rum* alone. "Prevention is better than cure." I am satisfied that musquitoes and gnats rarely trouble any one whose blood is not in a feverish and unhealthy state. Such a condition of the blood may result from sickness, but always follows the use of intoxicating alcoholic stimulants. I have fished from a canoe at night-fall, when these insects arose like clouds, apparently from the water, without material discomfort, while my companion suffered agonies. I told him (as a joke) it was because I was a radical and he a "copperhead." Your delicate, metropolitan dandy, who adores champagne suppers, and warms himself with brandy, had better keep clear of the North Woods. A person of frugal habit and diet can bear bites and wounds, which would become festering sores and gangrened ulcers upon the body of the intemperate. If a preparation is desired, I should substitute hard mutton-tallow for hog's lard in the pennyroyal ointment. Mutton-tallow is worthy of a word of praise; to suppress an itching, to cure a bite or a galled spot, where the cuticle has been rubbed off, it is really invaluable.

In case furs have to be *cached* they may be cased in a tin or sheet iron can, proof against small animals, and then put far beyond the reach of bears or wolverenes. This is a good way to dispose of them at any time.

You should patent some light machine for setting the springs of large traps, by lever or jack-screw.*

---

branches are cut from a tree and thrown over it. They fear a trap. The deer or moose may also be cut up, and the parts swung up on small trees. Bend down a sapling as stout as you can handle, cut off a limb, hang the meat to the hook, and let the tree swing back. It will be out of reach of the wolves, and the tree will be too small for bears to climb. Moose-wood bark makes a good substitute for a rope, — Editors.

* Such contrivances are cumbersome to the trapper. For setting large, double-spring traps, he should use double levers made of wood. All that is necessary to be carried into the woods to do this is four strong leathern straps furnished with buckles. When you wish to set a trap, cut four levers of a size and length proportioned to the size of the trap. Take two of them, make a loop of one of the straps and slip it over one end of each; then bring the trap spring between them, press them together and adjust a loop over the other ends of the levers. Serve the other spring in the same way. Now spread the jaws, adjust the dog and pan, loosen the levers and

A good sledge for hauling stuff over the crust or snow in winter should be six feet long, eighteen inches broad, and six or eight inches high; as light as possible, held by iron braces running over the top and down the sides; very lightly shod.*

I am sorry I have made this article so long, but the fact is, once started, I have found it hard to stop. I take much interest in trapping, and seldom am happier than when I traverse the wilderness in pursuit of fur. Your book has been a great treat to me. It fills an odd little corner in literature, which but for you, might ever have remained vacant.

F. R.

your trap is set. The straps weigh only a few ounces and are easily carried. — Editors.

* The Indian sledge is better. It is made of a smooth board six or eight feet long, and fifteen or twenty inches wide, bent up in a curve at the forward end. It is light, does not sink in the snow or cut the crust, and draws easily. — Editors.

PLAN OF A TRAPPING CAMPAIGN.

By PETER M. GUNTER.

I BEGIN a trapping campaign, by selecting my hunting ground, building my shanties, making my canoes, carrying my traps to proper localities, and carrying in provisions.

In selecting a trapping ground it is a great advantage to get where you can travel by water as much as possible. You are likely in that case to capture more mink and otter. I manage in this way: I take a trip in a circle, following lakes, rivers and small streams, and striking across from one to the other, till I come round to the starting point. At this point I build a wigwam. This I do in the following manner: I cut four crotches, each about six feet long, and sharpen their lower ends. I stick two of them into the ground eight feet apart. Then I place a pole four inches in diameter on the top. This forms a plate for one side of the building. Four feet distant, and parallel to these, I place the other two crotches with a similar plate. Then I place other poles across the ends from one plate to the other. This done, the frame of the wigwam is finished, ready to inclose. Now to do this with only an axe would bother many. I do it in this way: Fell a cedar or any other tree that splits free, and cut off logs about twelve feet long. Split these up into boards for the roof. Lay one end of the boards on the ground, the other on the plate. Cover both sides in this way. Thus your roof is finished, leaving a space about two feet wide along the peak for a chimney. Then split some more boards for the gable ends. These are short and may be placed in an upright position. The door may be a split board. It should be opposite the fire, and open to the north to prevent smoke. Fill the crevices with moss to keep the wind out, and the structure is finished. Build your fire in the centre; that makes a partition; you have one room for a sleeping apartment, the other for a dining-room. This is my home shanty. It is quite necessary to have other shanties on the trapping line, to stop in over night, as I always calculate to be three days going round a circle, in setting and tending traps.

What I call an outfit for a trapping campaign, or at least what I take, is, one large axe to the home shanty, where I do my cooking, a tin six quart pail, for carrying water and other purposes, a pint cup, a sheet-iron bake-pan with lid, for baking bread and cooking game in, and a blanket, leaving it at the home shanty. I always carry a gun, (and prefer a double barreled shot and rifle gun,) a small axe weighing ten or twelve ounces, a pocket knife, a butcher knife in my belt, and from eighty to one hundred and fifty traps for one line. If there are many beaver you want one or two traps to each family. Sometimes I use the No. 1 Newhouse trap with good success for otter and beaver; and I have caught four wolves in that sized trap on land. But I prefer for my own use, for taking beaver and otter, the No. 2 or fox trap. In the way of provisions, I carry butter and flour, and some tea, salt, and pepper. For meat I depend on my gun and traps.

In setting traps attention should be paid to the signs of game. These are well known to old trappers, and are learned by careful observation.

Beaver can easily be found in the fall by their cutting timber for their winter supply of food, and for repairing or building dams. During the summer they play about, laying up nothing, and feeding on aquatic plants till about the first of October. At this time, dam beaver begin to build their dams, and draw in timber for winter supplies. Bank beaver never build dams but live in the banks of streams, in holes lined with grass and leaves. Their holes start from the bottom of the stream, or at least three or four feet under water, rising up into the bank, above the level of the water, so that they are dry to sleep in. Bank beaver feed like other beaver, drawing sticks into their dwellings, eating the bark off, and then carrying the refuse out into the water again. In building their dams beaver always choose a location at the head of rapids, where they can have open water in winter. Bank beaver generally build their habitations along the sides of rapids.

Beavers in travelling on land generally have one particular path which they follow; therefore, if you set a trap at each end of the path you are quite sure to capture them. The trap should be set a little on one side of the middle of the path, and three or four inches under water. In a single trap, set in this way, I have caught two otters, four beavers, and seven muskrats, during one trapping season.

The otters' haunts are detected by their slides, and the freshness of their works on the slides.

Mink, marten and fisher, have no particular signs except their foot-prints and droppings, generally where they cross from one stream to another. Minks have certain run-ways the same as deer. On these run-ways they always stop in some old root or hollow log. When you find one of these places, you can tell whether it is a mink-haunt by their droppings. Set your trap in or near these holes and you are sure to catch any mink that passes. I have caught four minks in one season, in one hollow log, without using any bait. If there are deer run-ways on your hunting grounds, marten and fisher will follow those paths, in order to pick up provisions. In these places the wolf is the marten's and fisher's provider. Nearly all the deer that are killed by wolves, are killed on the run-ways, and the marten and fisher follow the wolf to pick up the fragments he leaves. Hence, whenever I cross a deer's run-way I set a trap or two, and generally with success.

During the last five years I have been trapping in partnership with Mr. Robert Holland, an accomplished deer-hunter and trapper, and by way of conclusion to this article I will give the results of our labors for three years. Our method is to carry on farming during the summer months, and trap in the fall, winter, and early spring. In 1863 we caught ninety-eight minks, fifty-two martens, fourteen fishers, ten otters, fifty-three beavers, five wolves, thirteen raccoons, seven foxes, and two hundred and eighty muskrats. In 1864 we caught eighty-nine minks, forty-seven martens, nine fishers, nine otters, ten foxes, six raccoons, two hundred and forty muskrats, five wolves, and sixty-two beavers. In 1865 we kept no account of the number of skins, but our sales amounted to $505. During these three years we caught one hundred and thirty-seven deer.

BOAT BUILDING.

By T. L. PITT.

A BOAT is often an indispensable part of the trapper's outfit. I will give a few general rules for the construction of the several kinds in use.

THE BARK CANOE.

This is the favorite boat in those regions where the canoe-birch grows to perfection. It is of Indian origin, and usually of Indian construction. Few white men are sufficiently versed in the art of making it to rival an experienced Indian in the nicety of work.

The great advantage of the bark canoe, or the "bark," as it is usually called, is its *lightness*. On this account it is preferred on all streams where portaging is necessary. A large sized one, fifteen to twenty feet long, may be carried without difficulty on the shoulders of two men; while a small one, ten or twelve feet in length, can be carried by one man. They are built of all sizes, from ten to thirty-five feet in length. The largest ones will carry a dozen persons or more, besides considerable freight.

In building a "bark," a cedar gunwale is first prepared. This should be composed of two strips for each side of the canoe, about one fourth of an inch thick, and an inch or more in width, one to go inside the edge and the other outside. The bark is then procured. That part which forms the bottom of the canoe should be in one whole piece, carefully peeled from a tree of suitable size and free from knots. If not large enough for the whole boat, strips may be sewed on to it. After the bark is ready, the length of the proposed

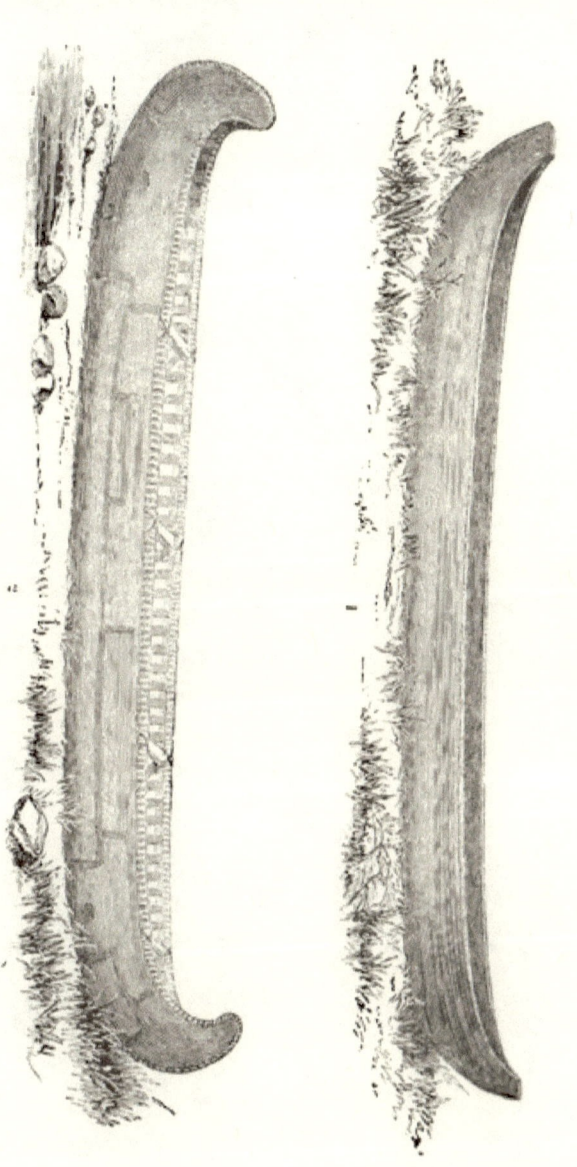

CANOES.—1. THE DUG-OUT.—2. THE BARK.

canoe is measured off on the ground, and at each end of the space two stakes are driven firmly into the earth, close together. The ends of the bark are then folded on the middle line, with the inside of the bark outward, and inserted between the stakes. These ends should extend beyond the stakes far enough to allow a strip of bark to be folded over them, and the whole firmly sewed together. This makes a rude form of the canoe. Underneath each end, near the stakes, a small log is placed, for the canoe to rest upon, and to let the bottom form an appropriate curve downwards. The gunwale is then placed in position, the bark fitted between the strips, and the whole sewed together with a winding stitch, regularly, or in sections, the entire length. Next the inside of the canoe is lined with strips of cedar, from one fourth to one half of an inch thick, and an inch or more wide, placed longitudinally and fastened in place with pine pitch. These strips may be several feet long, and should neatly lap where their ends meet. Knees or ribs are then made. These are strips of ash, or any wood that is firm and elastic, and should be about one fourth of an inch thick, and from one to two inches wide. They are placed crosswise of the canoe, bent down to the bottom and sides, and their ends securely fastened under the gunwales. They should be placed close together or with alternate spaces between them, the whole length of the canoe. They strengthen the canoe, keep it in shape, and keep the lining in its place. When all this is done, the whole inside of the canoe and all the seams are smeared with pitch, and two or three cross-pieces are placed between the gunwales to keep the sides in shape. The sewing is all done with a square or three-cornered awl, using fibrous cedar, spruce, or tamarack roots, soaked in hot water, for thread.

THE LOG CANOE OR DUG-OUT.

This is a kind of boat often built by the trapper. Its construction is simple; it may be made quite light; it is strong, serviceable, and durable. A log canoe may be made of pine, whitewood, butternut, black-ash, basswood, or cotton-wood. The best are made of pine. A log suitable for this purpose should be large, sound, and free from knots. It should first be hewn on two opposite sides to a size corresponding to the depth of the intended canoe. On one side the hewing should not be on a straight line, but should run out at the ends to the surface of the log, in order to leave a suitable rise at bow and stern. This hewing is usually performed before the log is cut off from the tree. When this is accomplished the log is turned down, with that side uppermost which is to form the gunwale. Next, the outlines of the sides are struck with a line and chalk, the latter being usually a burnt stick. The general rule for laying out a canoe, is to measure the log into three equal sections. The two end sections are for the bow and stern respectively. For a large canoe the bow should be hewn somewhat sharper than the stern. At the same time the width of the boat at the point where the curves of the bow start, below the gunwale, should be a little greater than at any other point. This difference can be easily attained in finishing off the sides, after the general shape is struck out. If the canoe is very large it may be desirable to attend to this point in the first hewing. The object in giving the canoe a greater width at this part is, to give ease of motion in the water. The same principle that governs in the construction of larger vessels, and is seen in the shape of the duck or goose, applies to the shaping of a large canoe. A small canoe, for running deer, and designed to never carry more than two persons, may be curved with the same sharpness at both ends, and have no variation in its width. It may then be run either end foremost. A canoe made in this way, if narrow and very sharp, in skillful hands, may be one of the swiftest and most effective boats. Both ends of a well-made canoe are curved upward from the middle of the gunwale, and the stern rises a little from the line of the bottom. When the tree is sound (and none other should be used), a canoe may be worked very thin, and thus be so light as to be easily carried. With all these points in mind the canoe is hewn to nearly its final outside shape; the inside is dug out with axes and an adze; finally it is neatly and smoothly finished — on the outside with axe and draw-shave, and on the inside with a round edged adze or howel. The tools required in making a log canoe are, a good common axe, a broad axe, a common adze, a howel or round adze, and a large draw-shave. A small auger is also desirable for gauging the thickness of the bottom by boring, and, if obtainable, a cross-cut saw saves labor.

SPRUCE BARK CANOES.

Rough, temporary canoes may be made of spruce or basswood bark, by simply folding the ends and sewing or nailing them together, adding gunwales and lining, putting in a few knees and cross-pieces, and smearing all the joints with pitch.

BATEAUX

Are made of thin boards, nailed together in the form of a flat-bottomed boat. Select two boards that are sound and free from knots, and of a length and width equal respectively to the length and depth of the proposed boat. Set the boards up edgewise, the width on the gunwale apart, and nail on a cross-piece midway between the ends. Then turn the boards over and, with a draw-shave, shape the other edges to a proper curve for the bottom. Next, nail a board across the middle of the bottom; then bring the ends of the boards together and nail them to the bow and stern pieces. The bottom is then made by nailing boards crosswise, care being taken to give the sides a proper curve. After all the parts are put together, the joints are caulked, and the bateau is then ready for use.

SNOW-SHOES.

The proper form of a snow-shoe and the mode of fastening it to the foot are shown in the illustration on the opposite page. The frame of the shoe should be made of ash or some other strong, elastic wood. The interlacing should be composed of strips of deer-skin, moose-skin, or untanned neat's hide. Two methods are followed in fastening the interlacing to the sides or bow of the shoe. In one case the bow is firmly and closely wound with strips of skin, and the interlacing is fastened into the winding. In the other case the winding is omitted and the interlacing is fastened through holes bored at regular intervals in the bow. Snow-shoes are indispensable to the trapper wherever deep snows prevail.

OIL FOR FIRE-ARMS.

The trapper should always be provided with oil for his guns. Probably the best kind he can use is purified neat's-foot oil. It is prepared in this way: Drop a few strips of lead or some shot into a bottle of the oil and then place it in the sun's rays. A heavy deposit will take place, filling the lower part of the bottle. The upper part becomes bright and limpid, and by a repetition of the process may be so effectually purified that it will never be liable to viscidity. It is in this manner that watchmakers purify the oil used in lubricating their delicate machinery. Oil prepared from the fat of the Ruffed Grouse is also good for fire-arms when the above cannot be obtained.

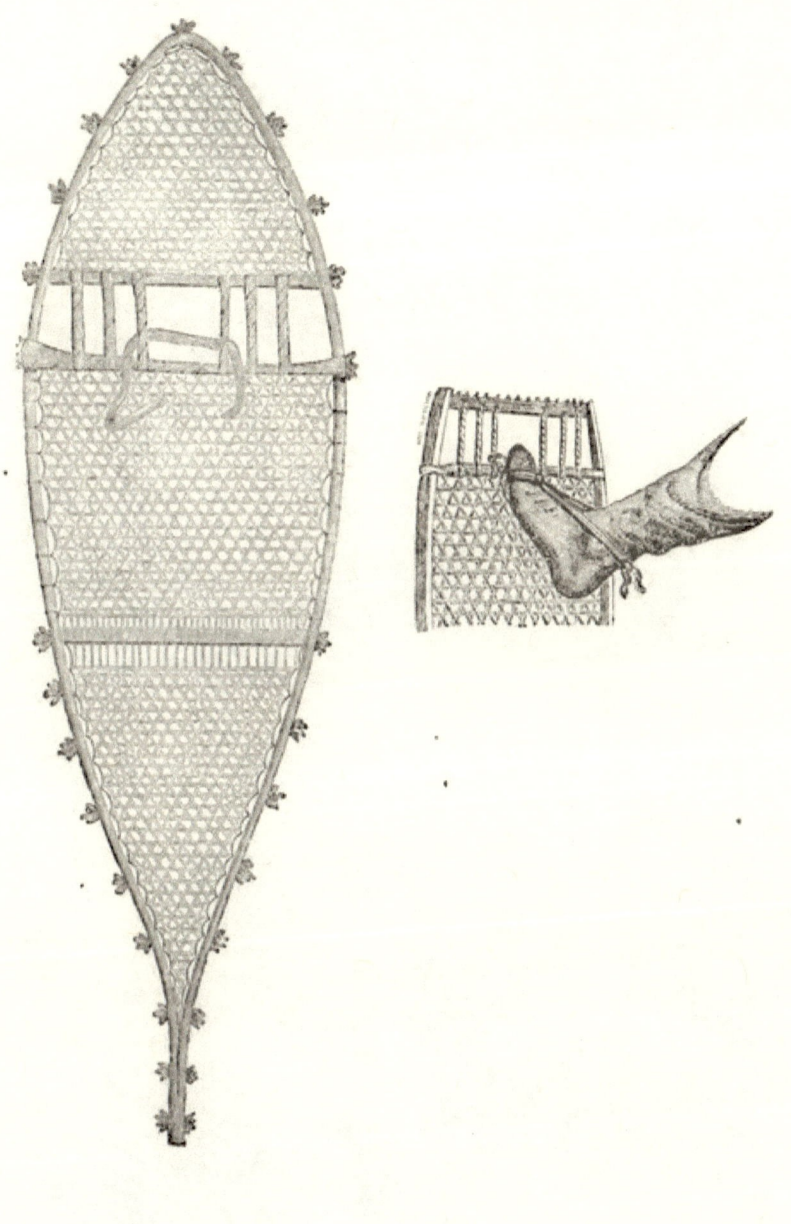

NARRATIVES.

[In the first three of the following articles illustrating the trapper's life, we introduce to our readers the Hutchins family, the father and two sons — a trio of "mighty hunters." — EDITORS.]

AN EVENING WITH AN OLD TRAPPER.

BY W. A. HINDS.

OF all story-tellers, give me those who have spent the greater portion of their lives in hunting, fishing, and trapping; who have lived for weeks on wild game; who have tramped for months alone through the forests; who have camped on green boughs, or kept themselves comfortable in deer-skins, when the thermometer was far below zero. Such men inspire me with a degree of respect like that entertained for all whose lives have been heroic. Soldiers of the woods, they have often endured hardships superior to those who have carried the knapsack in the open field. Though in many instances unfamiliar with books, they yet have a power of graphic and forcible description, seldom possessed by those who have made language their study. After conversing with them an hour, one feels as though he had himself encountered the bear and the panther, and been successful in hunting the otter and mink.

It would be difficult to find, at least in the Eastern and Middle States, a better representative of this class than Mr. John Hutchins, now a resident of Manlius, N. Y.

Born in Portland, Somerset County, Maine, November 16, 1801, he is consequently now (1865) nearly sixty-four years of age; but he is still "eager for the chase," and is planning a trapping expedition into Canada for the coming autumn. For more than half a century, he has spent a portion of each year in trapping and hunting. In his tenth year he caught and shot seventy-three squirrels, six blue jays, one mink, one weasel, and six partridges. When fourteen years of age he caught a bear which had killed a cow in the neighborhood where he lived in Maine; and he estimates the number of animals which he has caught in traps, or otherwise destroyed, as follows: 100 moose; 1000 deer; 10 caribou; 100 bears; 50 wolves; 500 foxes; 100 raccoons; 25 wild cats; 100 lynx; 150 otter; 600 beaver; 400 fishers; mink and marten by the thousands; muskrats by the ten thousands.

After reading the above list, no one will doubt his skill and wisdom in wood-craft, or question the probability of the adventures he relates. He is always ready to communicate to others what he has learned in his long life in the woods; and he takes the same pleasure in recounting his adventures that the scar-worn soldier takes in telling of battles, sieges, and marches. On meeting Mr. Hutchins a short time since, in company with his son, I interrogated him in true Yankee style, as follows:—

"In what part of the country have you trapped and hunted?"

"Mostly in Maine, Lower Canada, New Brunswick, and New York, but some in Vermont and in Michigan."

"At what seasons of the year do you generally trap?"

"I generally commence about the first of November, and trap till the first of April. There is no certainty of securing prime fur before the first of November, and but few kinds are good after the first of April. The three kinds—beaver, otter, and muskrat—are, however, good till the first of May; and the fur of the otter is good even as late in the season as June."

"Do you generally go alone, or with companions?"

"I have trapped alone about one fourth of the time. It is generally more pleasant, but less profitable, to have companions. When game is plenty, it answers well to have partners; but I would recommend never to have more than two, and think it nearly always better to have only one companion."

"How many animals have you generally taken on a winter's trip?"

"That depends, of course, entirely upon my fortune in securing good trapping ground. My son Samuel and I trapped one season in Upper Canada, and caught forty-seven beaver; and the furs of other animals, which we caught at the same time, would bring as much money as that of the beaver. The best specimen of luck I ever had was in setting twenty-seven traps, and finding a mink, fisher, or marten in twenty-five of them. That was on my second trip to Canada."

"How much money did you generally make?"

"That is another difficult question. I have made from $5 to $75 a month."

"Well, then, how much did you make in your best trip?"

"The best trip I ever made was forty years ago. I went out on Dead River, in the State of Maine. I was absent from home just one month (started December 3d, and returned January 3d); sold my fur for ninety-seven dollars, and fur was then very cheap. The same fur would now bring several hundred dollars. Two of us have often made $100 a month, or $50 apiece."

"What do you take for an outfit?"

"A double-barrel gun; a hatchet (I used to carry an axe, but now prefer the hatchet); a butcher-knife; a pocket-knife; a camp-kettle holding about six quarts; a frying-pan; a pint dipper or cup, and a spoon. I go lightly clad, never taking an overcoat, and only a single woolen blanket. For a winter's campaign, I take 40 lbs. flour, 10 lbs. pork, 6 qts. beans, 5 lbs. sugar, and 1 lb. of tea. The two last items might be dispensed with. I have lived a week at a time in the woods, eating nothing but moose meat; and Reuben Howard, a trapper from Connecticut, says he has lived two months at a time on deer's meat alone."

"If you were starting now, wouldn't you take some little conveniences for cooking and camping, beside those you have mentioned?"

"No; the longer one lives the life of a hunter and trapper,

the better he learns to get along with few conveniences, and the more desirous he becomes of avoiding luggage."

"How many traps do you take along?"

"When I first went trapping, I thought six or eight traps enough; but steel-traps are so much better, and more easily tended than wooden traps and dead-falls, that I now take one hundred muskrat or mink traps — sometimes even one hundred and fifty — besides a few otter traps, and, if I am going into a beaver country, a dozen beaver traps."

"But you can't take all these into the woods at once?"

"No; I first select my trapping ground, and then 'make a line,' as trappers say; *i. e.*, carry into the woods three or four back-loads of traps, and deposit them in safe places along the line on which I intend to trap, which sometimes extends from twenty to forty miles, from one stream to another, or from one lake to another."

"How many traps can one man tend?"

"That depends, of course, upon circumstances. Where game is plenty, fifty traps will keep you skinning and stretching; but in other places you might tend one hundred and fifty or even two hundred traps."

"How did you camp at night?"

"There is a good deal to be learned about camping out. When I go into the woods to trap for any length of time, I generally build a home-shanty of logs or bark. If I want to build one which will last three or four years, I make it of logs, notching or dovetailing the ends, and laying them up in blockhouse style, filling the cracks with moss, and making a roof of split cedar or bark. Sometimes I make a shanty by simply driving down two crotched sticks, placing a pole on them, and sticking down poles all around excepting in front, and covering them all over with spruce bark. When near the home-shanty I sleep there of course, but at other times I have no covering excepting a single blanket. I find a big log, and make my bed of boughs on that side of it least exposed to the wind. If the snow is deep, I select my camping-place on the hill-side, digging down to the ground to make a fire, and sleeping myself on the snow below, so that the blaze of the fire will shine directly upon me. When travelling by water, I draw the boat on to the bank at night, partly turn it up, and sleep under it, building a fire a few feet distant in front. I generally have slept very soundly in the woods."

"I have kept you answering questions a long time; but I shall not leave fully satisfied unless you will give me an account of some interesting adventures, of which you must have had many in your half-century's hunting and trapping."

"My experiences have not been so thrilling as those related in many books; besides, I am a poor hand to tell stories."

"Tell him how you once nearly froze to death," said his son John, always pleased to hear his father repeat his adventures.

"Well, then," replied Mr. Hutchins, who only waited for a little urging, "I will tell you of my

ADVENTURE ON THE DEAD RIVER.

"It must have taken place nearly forty years ago in the State of Maine. It was on my second long trapping expedition. I went into the woods with one Captain John Churchill, a great trapper and hunter. After we had killed nine moose, we concluded that one of us had better return home and notify our friends and neighbors that they could have plenty of moose meat by coming into the woods after it. And so I started home for that purpose. We were then on the headwaters of the Androscoggin, about thirty miles from the headwaters of the Dead River, where our home-shanty was. The plan was for me to follow our line of traps, taking along what fur I found, and skinning and stretching it at the home-shanty, where I was to remain the first night. But instead of doing so, I thought, on reaching the shanty, as the sun was still an hour and a half high, that I would leave the fur for Churchill to skin, and go on several miles further. It was fifteen miles down the Dead River to Folsom's house, but I thought I could go about half way, to the place where Captain Churchill and I had camped when we went into the woods. So I tramped on. It was one of the cold, sharp, biting days in February, and the wind blew and the snow flew awfully. I got to the shanty about dark, and carefully collected a pile of dry sticks for kindling, spread my blanket in the corner, and prepared to have a comfortable night of it. Then I went to my knapsack to get my flint and steel to light my fire with, but they were missing. I searched every corner in vain, and finally concluded that I had left them back in camp. By this time it was dark, and piercing cold, and I hardly knew what to do. It was too late to think of returning to the camp, and I knew I should freeze to death if I remained where I was. So, after thinking it all over, I concluded to go on to Folsom's. I thought if I could get to the river the ice would be strong enough to hold me, and it would be easier travelling, and a straight road.

"I continued my course down the river until I came to a series of falls. Here the river was open, and I was obliged to leave the ice and travel on land until I got by the falls. After, as I thought, I had got by all danger, and supposing the ice strong enough to bear me, I grasped an alder-bush and slid down on to the ice. But the ice wasn't as strong as I had calculated, and so, instead of landing on solid bottom, I went straight through. I went in up to my neck, and was only saved from going completely under by the alder-bush, to which I still clung. I managed by dint of some maneuvering to disengage myself from my snow-shoes and knapsack. These, with my hatchet, I shoved from me on the ice. I then pulled myself out and went on; but before I had gone twenty rods my clothes were frozen stiff. I kept on for some distance further down stream, to where the river was not so rapid, and concluded to try the ice again. But I had no better luck than before. The ice gave way, and in I went again, just as I did before. I felt pretty bad, I can tell you, about that time; but I managed to get out and go on again. The walking was so hard that I could n't help trying the ice once more. I ought to have known better, or at least taken better care, after getting in twice; but somehow I did n't. I slid down on to the ice, and in an instant found myself in a little worse situation than I had been before. The ice was nearly but not quite thick enough to bear me; and I was so far from shore this time that I could not pull myself out. I floundered about among the broken ice and water for quite a little while; but finally managed to relieve myself of my luggage somewhat after the same manner as before. I then succeeded in reaching the shore, not in very good trim for travel either, for the ice, which had frozen on my clothes during my three duckings, made them very stiff and heavy.

"In spite of all this I managed to get to Folsom's; but here I had another disappointment. No one was there, and the fire was all out. Of course I could not stop in the condition I was in, as I should have frozen to death in half an hour. The nearest house was at Reed's, fifteen miles further down the river, and there was no other way for me but to get there as soon as possible.

"So I started down the river for Reed's. It was eleven or twelve o'clock at night, and I had a pretty hard time of it, but got there at last. Reed's house was on a hill; and when I got to the foot of that great hill I could n't walk up it to save my life; I had to crawl up on my hands and knees. Finally I got to the house and rapped at the door, and Reed came and opened it. I suppose I did look rather forlorn; at any rate, he seemed almost frightened at first. 'For God's sake, Hutchins, is this you?' were the first words he said. I explained my circumstances to him, and he took me into the house, built up a big fire and thawed me out, and then put me to bed, where I slept till the next day at noon, and then got up, feeling as well as usual, only a little stiff.

"It was thirty miles from the place on the Androscoggin where I first started from, to our camp; fifteen miles from there to Folsom's; and fifteen miles from Folsom's to Reed's, — in all sixty miles. I started from the Androscoggin at eight o'clock in the morning, and got to Reed's at half past three the next morning, making the whole sixty miles in nineteen hours and a half. I think if I had allowed myself to be frightened or disheartened, I should have gone under; but I kept up good heart, and came out all right."

A YOUNG TRAPPER'S EXPERIENCE.

By JOHN P. HUTCHINS.*

My earliest recollections are of the forest. My father was an experienced hunter and trapper, and when I was but five years of age I accompanied him on one of his expeditions into the great Maine wilderness in search of game and fish. I have a dim recollection on that occasion of hooking on to a very large fish, and of being unable, with my slender strength, to get him into the boat in which I was seated. This childish disappointment made quite an impression upon me, and I used anxiously to look forward to the time when I should be a match for any of the beasts of the woods, or the fish in the waters.

I was sufficiently old to endure the hardships of forest life, when my father took up his abode on the southern border of the great New York forest, sometimes called "John Brown's Tract." There we prosecuted the business of trapping in earnest. We stretched a line of traps nearly forty miles in length directly into the heart of the wilderness, over rivers, mountains, lakes, and plains; and along this line we diligently trapped the otter, fisher, marten, mink, muskrat, and raccoon.

To give an idea of the management of a practical trapper in the woods, I will describe in detail the operations by which we subsisted, and took our game while in the woods.

As our line of traps was about forty miles in length, and of course involved a journey of eighty miles to and from our home, our outfit became at once a very important considera-

* A member of the Oneida Community.

tion. In the first place, we must have enough to eat, and the means wherewith to cook our food; and at the same time we must not overload ourselves with luggage, as every pound of our personal effects must be carried on our backs for long days, through a pathless wilderness. The object then was to secure the greatest amount of nutriment with the least possible weight.

And then, not only food, but other absolute necessaries must be provided. We must have the means for procuring fire, for securing game and fish, for taking and disposing of our furs, for keeping warm on a cold night, &c.; all of which weigh down seriously, but can by no means be overlooked or omitted.

I may as well here remark, that about one fifth of the luggage generally recommended by writers and book-makers who treat of life in the woods, as suitable for the hunter's or trapper's outfit, will cover all his absolute wants. The remaining four fifths the old woodsman will consider as luxuries, if not superfluities. I suppose that, as a general thing, writers are not practical hunters or trappers, and this may account for the discrepancy I have mentioned.

A trapper makes great account of his fire. Aside from its primary use in cooking his food, it oftentimes supplies the place of house and bedding. Some carry with them a light woolen blanket, but oftener the woodsman has only the earth for his resting-place, and the heavens for his counterpane, a sheltered nook, where the wind cannot blow too rudely, a few hemlock boughs for his bed, and a fire just in proportion to the temperature of the season.

Aside from the necessary supply of traps, the trapper's outfit can be reduced to about the following items:

First. A basket or knapsack, to carry on his back, and large enough to hold provisions and other necessaries for the journey.

Second. Eatables, consisting principally, or wholly, of pork and flour; or, what is better on some accounts, a mixture of flour and Indian meal, in the proportion of two parts flour to one of meal. Add to this a little saleratus and a small bag of salt, and a man can carry food sufficient, with what game and fish he can procure, to last him a month. It is much easier to carry the flour into the woods and bake it as it is wanted, than to attempt to use bread already baked, as it is lighter and less bulky. When the woodsman wishes for bread, he mixes the flour in a basin of warm water, adds a little saleratus and salt, and bakes it in his frying-pan, or if that is not at hand, on a chip.

Third. Cooking utensils, namely, a small frying-pan, two tin basins of the capacity of one and two quarts respectively, and a small tin cup for drinking.

Fourth. Implements for general use, namely, an axe, gun, knife, and pocket-compass.

Lastly, and above all, a good supply of matches.

Every trapper should have a companion to assist him, as the same gun, axe, and cooking utensils will suffice for both, and it is much less labor for two than for one to carry them.

When the business of trapping is prosecuted on the borders of lakes and large streams, much hard labor is saved by the use of a boat. Those who make free use of boats are more lavish in their outfit, as the labor of transportation is thereby very much reduced. I suspect that Mr. Newhouse has been more familiar with this method than myself; and this may account for any apparent discrepancy between us in respect to outfit.

When I began life as a trapper, I lived, as I have said, with my father, on the southern border of the great New York wilderness; so that our line of traps commenced not far from our home. This line was by degrees extended further and further into the forest, until it had reached the limit beyond which the provisions we could carry would not hold out. We began by carrying our traps into the woods, and distributing them along our intended line before the trapping season began; so that when the time arrived that fur was suitable for market, we should have only to set our traps and bait them. At the proper season we would shoulder our packs, containing as much provisions as we could comfortably carry, and commence carefully setting and baiting our traps. This process was continued as long as our provisions would allow, and then we would return on the same line, examining our traps, skinning the animals taken, and stretching their furs. After a short interval, this process was repeated, and kept up while the season lasted.

Our usual course was, to follow rivers and streams, and visit all the lakes in the vicinity of our line. When following streams, or the shores of the lakes, we would trap the beaver, otter, mink, and muskrat; and when our line extended over land and away from the water, we took the marten, fisher, and raccoon.

Our methods of setting and baiting traps, and our contrivances for circumventing animals, were generally very much like those recommended by Mr. Newhouse, and need not be detailed.

In the course of my trapping experience I had considerable practice in taking the fisher, and became somewhat familiar with its ways. This is a very pretty creature, with glossy black fur, and a long bushy tail. But, like the cat, it has a temper that is not so mild and agreeable as its appearance might indicate; nor does the close embrace of one of Newhouse's traps tend to mollify it at all. It frequently makes sad havoc with the trap and its appurtenances, and sometimes gets away after being fairly caught. I well remember a trying experience I had with one of these animals in the North Woods. I had seen his tracks, and had carefully set my trap with all the usual fixings and fastenings, in full faith in his ultimate capture. But on going to the place the next day, trap and chain were clean gone, and all fixings demolished. The fisher had been there, and had been caught, but instead of submitting handsomely to his fate, had gone and robbed me of a good Newhouse trap. (It was not Newhouse's fault.) He was a very large animal, and the spring-pole was not strong enough to swing him clear off the ground. So after demolishing the inclosure in which the trap was set, and making a general smash of things around, he threw himself upon the end of the pole, actually gnawed it off below where the ring was fastened, and left for parts unknown. How he finally disposed of the valuables he carried off, or whether he drew them about for the rest of his life, is left for conjecture.

I have long since abandoned the woods, and my trapper's life seems like a dream of the past; and yet I look back to it as a long and pleasant dream, despite of its many hardships and privations. In entering the woods I seemed to leave behind the jostlings and heartaches of crowded society — the great "*tom*" in which mankind are tumbling and chafing — and went forth into the freedom and peace of undisturbed Nature.

THE DEER HUNT.

FROM SAMUEL S. HUTCHINS'* JOURNAL.

OCT. 21, 1860. — We caught a deer to-day, and I am going to tell you all about it; for we had a lively time, I assure you.

It was one of those still, cloudy mornings you see so often at this time of year. We rose early, got our breakfast, did up our chores, and then started for the lake to hunt deer. We found the lake as calm and smooth as glass. Father took the large boat and went up to the head of the lake to start the dog, and I took the small boat and started down the lake for the "point," to watch for the deer. After getting there I climbed up into a tree, so that I could have a good view of the lake, and listened for the dog. After staying there some time, the wind began to rise, and I was cold, and began to think that we should hardly get a deer that day. So I came down out of the tree and begun stirring about to get warm, when I heard the dog away off on the hills. I stopped for a moment to see which way the chase was going, and came to the conclusion that they were coming around the head of the lake, and so on down to where I was. I then got up into the tree again, to await the result. I waited about an hour, I should think, watching the upper part of the lake most of the time, thinking the deer would be most likely to come in there. On looking, however, in the other direction, behold there was the deer, swimming for life. It was a buck, and a large one too. He was about half-way across the lake, and, half a mile from where I was. I did not stand there and look at him long, I

---

* This young man was a soldier in the late war, and came home from McClellan's peninsular campaign, with wounds and diseases that caused his death in the fall of 1864.

reckon. Down I came, twenty feet at two jumps, hurting my shins most wofully on the limbs, and my nose on the stones where I landed; but I picked myself up and got into my boat. Then commenced the chase. But let me describe the boat in which I was, so that you can better appreciate the fun. It is just eleven feet long, and sixteen inches wide, and scarcely heavier than an egg-shell, (poetic license,) and will upset a great deal easier. It was made from a bass-wood log, and well made too, and is what is commonly called a "dug-out." I had to stand on my knees in the middle, and had a double paddle, which is just like a common one, only it has a blade on each end. Thus equipped I started the chase, with the wind in my favor, and with the firm intention of catching the buck if I possibly could. He was a good half mile ahead of me, and had not so far as that to go to get to shore; and I could see that he swam furiously. I had no weapons to slay him with. My duty was to get around him, and drive him up the lake to father, who, when he saw me start out, I expected, would come and meet me and help kill him. So away I went, exerting every nerve and muscle; shot around the point, and was out at sea in "no time;" kept my eye on the deer, and took a course that would cut him off from the shore that he was swimming for. For a long while I went thus, with the wind in my favor, sometimes thinking that I should overhaul him, and then again that I should not. Finally I saw that I was gaining on him a little; but I knew that I must do more than that, if I wanted to catch him; so I redoubled my efforts. "Pull, Sam!" I muttered, "you must overhaul him, anyhow;" and so I did. After a long and hard pull I came up to him. When he saw me he turned square off from me, and swam almost as fast again as he did before. When I came about, side to the wind, to follow him, my little boat dipped water at every wave. But I stopped not for that. I wanted to run in beside him once more, and turn him toward the opposite shore; but I found that it was somewhat harder to do so than I expected. I laid out all my strength. You could have heard me puff half a mile off, if you had been within that distance. I could see that I gained on him, but very slowly. He sees that I am coming too near him, and he makes a short turn and swims for the middle of the lake — just where I wanted him to go, exactly! When I found he was safe, I dropped my paddle and shouted lustily for joy. Father came in a few minutes, and dispatched him, but not without a desperate battle. He fired three charges of buck-shot into his head, struck him more than forty blows with a hatchet, and only succeeded in killing him by getting hold of his legs separately and hamstringing him, after which he could raise his head sufficiently to cut his throat. He was an old buck of the toughest kind, and weighed three hundred pounds.

MUSKRAT HUNTING.

By HENRY THACKER.*

In the winter of 1844-5, I made two or three excursions from the city of Chicago into the neighboring wild regions for the purpose of spearing and trapping muskrats. At this distance of time I shall hardly be able to give from memory a very accurate account of those excursions; but I enjoyed them so well, and they made such vivid impressions on my mind, that I can at least give an outline of them, and shall recall as I proceed many interesting incidents.

The first thing I did, by way of preparation for the campaign, was to procure a suitable spear, which was simply a rod of round steel, three eighths of an inch in diameter, and three feet long, nicely pointed and polished at one end, and at the other driven firmly into a ferruled wooden handle, also about three feet long. The next thing (and a very important one) was to provide a pair of mufflers, made of old carpeting, to be drawn on over my boots. Lastly I harnessed myself into a knapsack suitable for carrying provisions, game, &c. Thus equipped, I put on my skates one morning, as soon as I found the ice strong enough to bear me, and started up the north branch of the Chicago River for Mud Lake, a small sheet of water about twelve miles distant, surrounded by extensive marshes, a noted place, not only for the habitation of the muskrat and mink, but for the gathering in the spring and fall of the year, of multitudes of almost every variety of wild ducks, geese, and other water-fowl.

Here let me describe the character and situation of this

* A member of the Oneida Community.

marsh and lake. The lake proper is a narrow sheet of water, from ten to twenty-five rods wide, and two or three miles in length. The water is from three to ten feet deep, and the soft mud at the bottom probably a great deal deeper. This lake seems to have two outlets flowing in opposite directions; one toward Chicago, being the principal head-waters of the south branch of the river which forms the harbor of Chicago; the other in the opposite direction, emptying into the Oplain River, which is among the head-waters of the Illinois River. I was told that at the time of the high water in June of that year (1844), schooners from Lake Michigan could easily have passed through this lake and marsh, into the Oplain, and so down the Illinois River to the Mississippi.

But to return to my story: on arriving at the marsh I found the ice strong enough to bear my weight, and quite transparent. A sight was here presented that I had never seen before. I cannot describe the view better than by likening it to a large meadow covered with hay-cocks, so thickly was the marsh before me studded with muskrat houses.

These structures are built up of flag-tops, roots, mud, and sea-weed, or water-grass, to the height and size of a hay-cock; and in them the muskrats live through the winter and spring. They generally commence their houses on a place where the water is one or two feet deep, and build it up entirely solid, to the height of three to five feet above the water, cutting out channels diverging in different directions from the house, and using the materials thus displaced in strengthening the foundation of the house. These channels are used as runways by the rats, in going back and forth between the house and their feeding-beds, during winter. After the superstructure is finished a hole is cut from underneath, up into the centre of the house, forming a nest just above the water, leaving ample room for a second story in case of a flood.

I now made preparation to enter upon the business of my excursion, that of spearing muskrats. I was not long in putting on my mufflers and getting ready for the onslaught; and, as this was my first attempt at spearing, I was full of enthusiasm. With feelings of interest and excitement, I marched up to a large house very cautiously (for, with the least jar or crack of the ice, away goes your game), and, with uplifted spear, made ready for a thrust. I hesitated. There was a difficulty I had not taken into account; I knew not where to strike. The chances of missing the game were apparent, but there was no time to be lost; so bang! went the spear into a hard, frozen mass, penetrating it not more than three or four inches, and away went the game in every direction. With feelings of some chagrin I withdrew my spear, and began feeling about for a more vulnerable spot, which I was not long in detecting. It being a cold, freezing day, I discovered an accumulation of white frost on a certain spot of the house, and putting my spear on the place I found it readily entered. The mystery was solved at once; this frost on the outside of the house was caused by the breath and heat of the animals immediately beneath it, and it was generally on the southeast side of the centre of the house, this being the warmest side. Acting on these discoveries, I made another trial, and was successful; and now the sport began in good earnest. Whenever I made a successful thrust, I would cut a hole through the wall of the house with my hatchet, and take out the game, close up the hole, and start for another house. The remaining members of the family would soon return, and immediately set about repairing the breach. I sometimes succeeded in pinning two rats at one thrust. I also became quite expert in taking game in another way, as follows: Whenever I made an unsuccessful thrust into a house, the rats would dive into the water through their paths or run-ways, and disappear in all directions. I now found I could easily drive my one-tined spear through the ice two inches thick, and pin a rat with considerable certainty, which very much increased the sport, and I was not long in securing a pile of fifteen or twenty rats.

Here I made a discovery of what, until now, had been a mystery to me, namely, how a muskrat managed to remain so long a time in the water under the ice without drowning. The muskrat, I perceived, on leaving his house inhaled a full breath, and would then stay under water as long as he could without breathing; when he would rise up with his nose against the ice, and breathe out his breath, which seemed to displace the water, forming a bubble. I could distinctly see him breathe this bubble in and out several times, and then dive again. In this way I have chased them about under the ice for some time before capturing them. I do not know how long the muskrat could live under the ice, but I have heard of their having been seen crossing large bays and rivers under the ice, five miles from shore. I saw a man in Illinois who told me he chased two otters under the ice for three quarters of an hour, trying to kill them with his axe, and finally lost them; which goes to show that these animals, as well as the muskrat, can live under the ice a long time.

As I frequently speared the muskrat on his feeding-bed, and subsequently found it to be the best and surest place to set a trap for him, I will, for the benefit of the novice, undertake to describe one as found in the marshes. A feeding-bed is a place where the muskrat goes to feed, generally at night, and is frequently many rods from his house. Here he selects a place where his food is convenient, and by the aid of the refuse material of the roots, &c., which he carries here for food, he elevates himself partly out of water, in a sort of hut. Here he sits and eats his food, and at the slightest noise, or least appearance of danger, disappears in an instant under water. In the winter these feeding-places are readily discovered by a bunch of wadded grass, flag, or some other material, about the size of a man's hat, protruding above the ice. This little mound is hollow, and is only large enough for a single rat, where he sits and eats his food, with his lower parts in the water. When the rats were disturbed in their house, I found they generally fled to these feeding-huts, where they were almost a certain mark for the spearman.

Finding I had taken as many rats as I could conveniently strip before they became frozen, I set about the work of skinning, and after an hour and a half of pretty cold work, I bagged my skins, put on my skates, and started for the city, well satisfied with my first day's excursion.

In my next excursion, not many days after, to the same

place, I had still better success. As the ice had now become too thick to be easily penetrated by my spear, I adopted, in part, a different mode of taking the game. This time I carried with me, in addition to my spear, two dozen steel-traps, and a bundle of willow sticks (cut on the way) about three feet long. On arriving at the hunting grounds I prepared myself for the day's sport by putting on my mufflers, and with traps and willow sticks slung upon my back, began the work by driving my spear into the first house I came to. I could not now see the rats as they fled from the house, on account of the thickness of the ice and a slight snow that lay upon it. Consequently the sport of spearing them through the ice was cut off. But as often as I had occasion to cut through the walls of the house to take out my game, I set a steel-trap in the nest, slipped a willow stick through the ring of the chain, laid it across the hole, slightly stopped it up, and then passed on to the next house; and so on, until my traps were all gone. I then started back to the place of beginning, driving my spear into every feeding-hut in my course, and killing many rats. Finally, I began going over the ground again, first driving my spear into a house, then examining the trap, taking out the game, and re-setting the trap. In this course I was quite successful. I found by setting the trap in the right place, near the edge, and a little under the water, I was almost certain to take the first rat that returned. In making two or three rounds in this way, I found the rats became somewhat disturbed, and sought temporary shelter elsewhere; when I would move to a new place, giving them time to recover from their fright.

I think this a very profitable method of trapping the muskrat, especially in an open winter. It very much lengthens the season of trapping, which is quite an important consideration with the trapper. Another consideration is, the trapper may set his traps and allow them to remain many days, if not convenient to go to them, and be sure his fur will take no harm; as the rat on being caught in the trap dives into the water, and is soon drowned, and will not spoil for a long time at this season of the year, and is also secure from frost.

I will here state that I found a muskrat house to contain from four to nine rats. I have caught as many as nine from one house. Possibly some may contain a greater number than this. I concluded that these colonies must be the progeny of a single rat in one season, or for aught I know, at a single litter.

In these winter excursions, I sometimes captured several minks, which I found somewhat different from the mink of the Eastern States, being much larger, and of a lighter brown color and coarser fur. I sometimes found them occupying muskrat houses, from which they had driven or destroyed the muskrats, of the flesh of which they are very fond. They are a gross-feeding, carnivorous animal. I have found stored up in muskrat houses which they inhabited, from a peck to half a bushel of fish, in all stages of decay, and some freshly caught and alive; which is good evidence that they are not only gross feeders, but good fishers also. I was most successful in taking the mink in steel-traps, baiting with muskrat-flesh or fish, and setting my traps about the marshes, and along the banks of streams and rivers. A mink will seldom pass a bait without taking or smelling at it; and by placing the bait a little beyond the trap, in such a position that he must pass over the trap in order to reach it, you are pretty sure of him. I also caught them by setting the trap in the mouth of their dens and in hollow logs, and sometimes enjoyed the sport of digging them out of the river-bank.

In setting my traps for mink and raccoon, I was somewhat annoyed by the prairie wolf taking the bait, but still more by the skunks getting into the traps. The country at this time abounded with these animals. They seemed to be nearly as plenty as the minks. I have sometimes found as many as two or three in my traps on a morning. It was an easy matter enough to dispatch one, but to do it and not get my trap scented was not so easy. (Here let me say, I never knew one caught in a trap to discharge at all, until disturbed by the approach of man.) After trying several unsuccessful plans, I hit upon one that I thought would do the business. Putting a tremendous charge of powder and ball into my rifle,

I approached my antagonist as near as I could without drawing his fire; and placing the muzzle of my rifle within three feet of his head, blazed away, and blew his head clean off. I approached the carcass for the purpose of taking off my trap, (congratulating myself on my good success), when he made a sudden convulsive movement, and, oh horror! such a discharge of the genuine article, no man ever saw or smelt! However, by a quick movement I escaped the charge myself, but my trap, as usual, was thoroughly perfumed. I soon had an opportunity to try again, and this time I succeeded, by the following device. Watching my opportunity when the skunk turned his eyes from me, I dealt him a heavy blow across the back with a long club, and immediately loosened the trap from off his leg. In this way I ever after managed to keep clear from scent, with a single exception, which occurred as follows: —

In one of my excursions, accompanied by another person, the dog scented something under the floor of an old shanty, which we concluded must be a mink; so at it we went tearing up the floor, to give the dog a chance to get at the animal. Up came one plank after another in quick succession, when all at once the dog made a tremendous lunge right into the midst of a nest of seven nearly full grown skunks. In less than a minute the atmosphere was blue with the most horrible stench ever encountered by human olfactories. The dog was soon nearly choked and blinded by the showers of stifling spray that met him at every charge, and, for the time being, all were obliged to beat a hasty retreat into the open air. But as we were all now fairly in for it, we concluded to make another charge and finish up the work we had so enthusiastically begun; and, armed each with a long club, we returned to the fray, and, with the help of the dog, soon despatched the foe, and retreated to the windward to get clear of the stench. But it was of no use. I seemed to be scented through and through; my very breath seemed to be hot with the terrible miasma; and for several days I could scarcely taste or smell any thing but skunk. This was my most serious encounter with the skunk family, though I continued to be annoyed by their getting into my traps; and once, at the suggestion of a fur-dealer that their skins were worth fifty cents apiece, undertook the job of saving a lot; but after skinning five, gave up the business in disgust.

My next excursion was a short but rather exciting one. In consequence of a slight thaw a day or two previous to my setting out, the skating on the river was nearly spoiled. I was therefore obliged to travel most of the way on land, and on foot, taking nearly all day to get to my place of destination. I put up for the night at a tavern a mile or two from the part of the marsh where I intended to trap, which was at the end opposite to the theatre of my previous excursions, and near the Oplain River. The next morning, after breakfast, I started out for the hunt, and, on arriving at the marsh, to my surprise not a muskrat house could be seen, with the exception of the very tops of three or four. The rest were all under water and the water frozen over. At first I was unable to divine the cause of this unusual rise in the water; but subsequently ascertained that an ice-dam had formed in the river three fourths of a mile below, in consequence of the breaking up of the ice above, and had set the water back over this part of the marsh to the depth of nearly four feet. The muskrats were completely drowned out; and I now saw them huddled together in numerous squads upon the newly-formed ice all over the marsh, having already brought up portions of their submerged dwellings, with which they had built up slight walls to shelter themselves from the cold northwest wind.

This was an exciting scene to the trapper — a multitude of his game in full view! I became almost nervous with excitement. But how to get at them was the question. On going down to the water, I found it scarcely frozen along the shore, though it looked firmer farther out. To be sure I could reach many of the muskrats with my rifle; but what was the use, if I could not get them after I had killed them? However, something must be done. I could n't stand this sight anyhow. I set about devising some plan by which I might reach the game in person. A half dozen plans were presented to my mind in as many minutes. One plan was to place a board on the ice, get on it, and shove myself along by placing the point of my sharp spear on the ice; but, on further consideration, I concluded this would be too slow an operation. If I succeeded in getting out on the ice, the rats could easily keep out of my way, as I should not be able to leave my board. Another plan was to fasten a piece of board a foot square to each foot; but, on further thought, this plan was also abandoned as being unsafe. Although the water did not exceed four feet in depth down to the old ice, yet, in case I broke through, the boards might operate to keep my heels up and my head down. I now determined to test the real strength of the ice; and, procuring a piece of slab twelve or fourteen feet long, I shoved it off on the ice. Leaving one end resting on the shore and walking out on this, I stepped off upon the ice. It barely held my weight, and soon began to settle, so that the water came upon the ice. However, I came to the conclusion that if I could get upon the ice with my skates on and keep constantly under pretty good headway, it would hold me up. Stripping off all extra clothing, and laying aside every unnecessary weight, I strapped on my skates, and, with spear in hand, launched forth in pursuit of the game. The ice bent and waved before me; but I glided swiftly on, and in less than a minute was among the muskrats.

I now discovered that the rats kept a hole open through the ice, right above their house; and, before I got within striking distance, they dove into the water and disappeared. I could hear them snuffing up against the ice, but could not see them on account of a slight sprinkling of snow which covered the ice. As soon as I left for another place, they would come up again through the holes on the ice. I saw that, in order to get a chance to strike them, I must wait at the holes for them to return for a fresh supply of air. This I found rather tedious, as I was obliged to keep constantly in motion, running in a circuit around the hole, on account of the weakness of the ice. In this way I would have to wait several minutes, and, when one did return to breathe, he was so very quick I found it difficult to hit him; and I also found, where the holes were not a great way apart, that when I went to one hole the rats would dive and swim to another. This would not do. I must try another expedient; and, returning to the shore, I took from my knapsack a dozen steel-traps and a handful of willow sticks, threw them on the ice, and then started back. Picking up in my course as many traps and sticks as I could carry without increasing my weight too much, I distributed them around the holes. And now lively work commenced. Taking a trap and stick in my hand, while under headway, I set the trap, slipped the willow stick through the ring of the chain, dropped it on the ice, placed the trap in the little cuddy where the rats huddled together, and passed on to the next, scarcely making a stop. This plan was a successful one. Frequently, before I reached the next hole, a rat would be caught in the trap I had just left, and, diving into the water, would be brought up at the length of the chain by the stick sliding across the hole, and in this condition would soon drown himself. I now had as much business as I could attend to, taking out the game, re-setting my dozen traps, carrying the game to the land, &c. You may be sure I played back and forth in a lively manner. I however discovered that the ice became much weakened by passing over it several times. Consequently I was under the necessity of moving to new places occasionally, to avoid breaking through. In fact, I found there was only a small part of the marsh where the ice was sufficiently strong to hold me up at all; and the weather, moderating after the middle of the day, weakened the ice so much that I fell through several times, getting my clothes wet and boots full of water; which so much increased my weight that I was soon obliged to abandon the field altogether. I had, however, by this time secured a good pile of rats, and, on the whole, had had one of the most exciting day's sport I ever enjoyed.

The weather now continued to moderate, and there were evident signs of the breaking up of winter, and the opening of spring. In two or three days from this time, wild ducks and geese began to gather about the marshes. I now began active preparations for a spring's campaign of trapping. During the winter two small trapping boats had been made, and a tent, camp-kettles, and other "fixings" had been got in readiness; and on about the twentieth of February, in company with E——, I set out. We launched our little crafts and commenced the campaign by scattering over the marsh one hundred and ten steel-traps, with open jaws, ready for the fur-clad inhabitants. The weather being favorable and the water steady, we made havoc among the muskrats and minks; and as this was a noted place for game, especially for muskrat, mink, and raccoon, we soon had competition in the business. In the course of three or four days, three other trappers stopped in the same vicinity, and commenced operations. But as they were strangers from a distance, we had decidedly the advantage, as we understood the ground, having previously pretty thoroughly reconnoitered the marshes in this section. The game being plenty, we found work enough to keep us busy, and for several succeeding days caught more rats than we could find time through the day to skin.

However, our good success was of comparatively short duration. In the course of ten or fifteen days, we found ourselves confronted by a pretty serious difficulty in the way of successful operations. As the previous summer had been remarkable for its long continuous rains and great flood, we now had the opposite state of things — continuous dry weather; and having had scarcely any rain the fall previous, nor snow during the winter, spring found the water in the rivers and marshes unusually low. As the weather continued fair, the March winds dried up the marshes so fast, that we soon found it difficult to get around with our boats, and finally were obliged to leave them altogether and take to the rivers, in order to continue our sport. We now found our chance for sport much reduced. The high water the previous spring and summer, overflowing the river-banks for so long a time, either prevented the rats breeding, or drowned their young, so that we found the game rather scarce. We however ascended the Oplain River some twenty or thirty miles. Our way was to string our traps along the banks, three or four miles at a setting, and then return to camp. The next day we would overhaul and re-set, if we found the game plenty enough to warrant it. If not, we would take up the traps and make another stretch, and so on.

On returning several days subsequently to our old hunting-grounds, we found the muskrats had somewhat recovered from the fright we had given them by our sudden and terrible onslaught, and had returned from the inaccessible parts of the marsh to which they had fled for refuge, and we made several more successful sets.

The weather had now become mild, and the marshes literally swarmed with ducks, and geese, and other water-fowl. Any one not familiar with this section of country can have no idea of the numbers of water-fowl that gather about these lakes and marshes in the spring and fall of the year. As we moved about in our little boats among the tall reeds and flags of the marsh, our fire-arms were always at hand, ready to bring down a duck or a goose that happened to pass within reach. We fared sumptuously every day. Our daily bill of fare consisted of roast goose, roast duck, prairie chicken, plover, pike, bass, cat-fish, bull-heads, &c., &c., together with coffee, hard biscuit, butter, and occasionally a meal of duck and goose eggs. This was what we called high living; and as we seldom found time for more than two meals a day, we were prepared to dispatch them with a relish that no one but a trapper can realize.

E—— did not seem to enter into the business with as much enthusiasm as myself, and having a family in the city, frequently found occasion to go home, and sometimes staid away two or three days. This made the work not quite so pleasant for me, as I enjoyed the sport much better when we were together. However, I got along very well; and the croaking of frogs, the peeping of lizards, quacking of ducks and geese, crowing of prairie chickens, the loud cries of the great sandhill cranes, and the almost incessant howling and yelping of prairie wolves, were all music to my ears. On the whole, I enjoyed the situation exceedingly.

One day as I was pushing my little boat along through the

tall reeds, I saw at a distance something unusual on the top of a muskrat house. As it was lying flat, almost hidden from view, I at first sight took it to be an otter, as we had killed one some time previous near the same place. As usual at the sight of game, my rifle was quick as thought brought to bear, and away sped the bullet, and over tumbled a large wild goose, making a great splashing as she fell into the water. On examination I found she had a nest of seven eggs, all fresh. The goose weighed fourteen pounds and a half. The same day I found another nest with several eggs, and took them to a farmer who was anxious to get them to hatch "at the halves." He placed the eggs under a hen; but a few days before they were ready to hatch, my ever-present enemy, the skunk, ate up hen, eggs, and all, to the great sorrow and indignation of the farmer. He said the young geese would have been worth five dollars a pair.

The weather still continued dry, and as we did not find game very plenty in the rivers, we concluded to wind up the trapping business, after having spent about six weeks in steady employment. We now collected our furs, and found we had caught seven hundred muskrats, sixty minks, a number of raccoons, and one otter; for which we found a ready market at good prices. Thus ended my first, and most interesting trapping campaign.

AN AMATEUR IN THE NORTH WOODS.

By CHARLES S. JOSLYN.*

It was a pleasant June evening when I first approached the southern boundary of the great New York wilderness. I had been an amateur sportsman from my earliest youth; and my fondness for the woods was, and has always been, quite inexpugnable. My feelings, therefore, when I came in full view of the long, dark line of primitive forest in the distance, were so exhilarating as to require some vent.

"Farewell, vain world!" said I, unconsciously breaking into a sort of monologue; "adieu to the pomp and glitter and artificiality of the thing they call society! Welcome, Nature, pure and unadulterated, fresh from the hand of the Creator!"

I was here interrupted by a smothered laugh from my companion, who had overheard the close of this rhapsody, which, in the exuberance of my feelings, I had uttered in a more elevated tone. Sewall Newhouse was a practiced woodsman, keen and shrewd, and well versed in the lore of the forest, but without much imagination or poetry in his composition.

"Wait awhile," said he, in his peculiar, dry way. "Don't be in a hurry about these things. Perhaps you will find some things in 'John Brown's Tract' that you don't calculate on. Besides, as it is getting dark, and we are several miles from the woods, we shall have to get one more night's lodging out of 'society,' as you call it, before we say good by to it."

The force of the latter consideration was quite irresistible, and had the immediate effect to postpone my enthusiasm for the time.

* A member of the Oneida Community.

It was after nightfall, when we succeeded in obtaining a lodging in the loft of a dilapidated farm-house, whose owner reluctantly consented to receive us. The accommodations were none of the choicest, and, accustomed as I was to clean sheets, soft beds, and other amenities of civilization, the general slovenliness of our dormitory, and the unyielding nature of our couch, were not at all conducive to repose. Newhouse, however, manifested an exemplary stoicism, and consoled me with the assurance that this was but a foretaste of what was in store for us.

The meagre amount of sleep which I enjoyed, and the general uncomfortableness of my surroundings, were favorable to early rising; and so we began our march soon after daylight the next morning. Our baggage had been sent ahead on horseback, so that we had but our guns to carry; and in the freshness of early morning, the hour's walk which brought us to the border of the woods seemed a brief one. A fence built directly across our path announced that we had reached the verge of civilization; and climbing this, in another moment we were within the precincts of the forest.

My first sensation was that of sublimity. An intense thrill of delight pervaded my whole being, and I almost involuntarily commenced repeating the opening stanzas of "Evangeline:"

"This is the forest primeval," &c.

My second sensation, which almost instantly dissipated the first, was that of mosquitoes — not the comparatively mild and inoffensive insect of polite society, but the savage and blood-thirsty vampire of the North Woods. Most of us have had experience with mosquitoes, and are more or less acquainted with the nature of the insect; but the mosquito of civilization no more resembles the mosquito of John Brown's Tract, than the bear trained to waltz to the music of the hurdy-gurdy resembles the untamed grizzly of the Sierra Nevada.

But thanks to the providence of my companion, help was at hand. Mosquitoes have an invincible repugnance to certain vegetable scents, the chief among which is, perhaps, that of pennyroyal. To prepare it for use, it is necessary to melt a certain quantity of lard, and add to it in its liquid state enough of the essence to infuse the mass with a strong scent. This compound, when cool, may be carried in the pocket in a tin box, and is an effectual preventive against the attacks of nearly every kind of insect peculiar to the American woods.

With this composition I plentifully anointed every visible portion of my body. Face, hands, ears, neck, every inch of surface which was liable to attack, was thoroughly lubricated, till I looked like an Esquimau just arisen from his dinner of seal's blubber and train-oil. The remedy, however, was effectual. It afforded me infinite satisfaction to see the impotent rage with which my late tormentors whirled round and round my head, in bewildering circles, never daring, however, to approach within reach of the aroma of this potent ointment. I anointed my face and neck twice or thrice a day, and found the application sufficient. The hands, owing to the necessity of use, require to be anointed about once an hour, to render them absolutely invulnerable. I found this somewhat tiresome, and subsequently adopted a pair of light buckskin gloves, which were not burdensome, and proved entirely mosquito proof.

In one of my excursions I met a young man who had incautiously ventured into the woods without adequate protection against mosquitoes. The blood was streaming from his face, where he had been bitten, and his general aspect was so forlorn that I was moved to pity. I gave him some ointment with directions how to use it, and left him. When I met him a few hours afterward, his first salutation was: "Mister, you've saved my life." The backwoodsmen become so accustomed to these insects, that they pay but little attention to them, in most cases using no defense against them. It is said that a mosquito will not bite an old hunter; and it is certain that after one has been in the woods a short time, these insects will pay much less attention to him than to a new-comer.

Mosquitoes however are not the only troublesome insect in

the woods. A small, black gnat, which the old inhabitants term a "punkey," bears away the palm from the mosquito. As these insects are only about one fourth as large as mosquitoes, they can penetrate the meshes of any mosquito-net; and when once they get scent of you, they will leave no portion of your body unexplored. The bite of these gnats is much worse than that of the mosquito. If you are bitten by the latter insect, and you do not unnecessarily irritate the wound, the effect is not visible for any great length of time afterward: but the bite of these gnats results, first in a deep crimson blotch about the size and shape of a half dime, and then in an open sore, which in some cases will last for weeks. The favorite method of protection against these insects in use in the North Woods, is, to build a fire with some damp material which will produce a dense smoke, plant yourself resolutely where the smoke is thickest, and take your chance of being smothered, as a choice of evils. Neither mosquitoes nor gnats can endure smoke; and this fact is taken advantage of by families living near the edge of the forest, who, during warm weather, keep a pan of embers continually smouldering at the doors of their houses, by way of self-defense.

Eight or ten minutes of brisk walking brought us to a small clearing, wherein an enterprising pioneer had constructed a rough dwelling, and ministered thence to the wants and necessities of incoming and outgoing travellers. The principal of these wants, I soon found, was whiskey. It is difficult for me to do adequate justice to this beverage. I am undecided, to this day, which of these two characteristic institutions of the North Woods is the worst, the whiskey or the mosquitoes. The rule is, I believe, that any one who can drink the whiskey can endure the mosquitoes; and, *vice versa*, any one who can endure the mosquitoes can drink the whiskey. Nevertheless, the article is in great demand, and indeed it seemed to be the common understanding that it was well-nigh impossible to undergo a two or three weeks' campaign in the woods without it.

It was eleven miles, said our informants, to our next stopping-place; and on we pushed, full of courageous intent, and bidding defiance to the perils and hardships of the wilderness. But the miles were unconscionably long. I had prided myself somewhat on my ability as a pedestrian, and had thought lightly of the eleven miles before us; but by the time we had accomplished one half of them, it seemed to me that each mile was a league in length. And then the path — how shall I describe it? I had thought the road by which we had reached the clearing in our rear as bad as road could be; but this path which we were now following was yet worse. If the reader will imagine an almost unlimited amount of logs, rocks, mud, stumps, and mosquitoes, mixed together hap-hazard, and distributed miscellaneously along a line eleven miles in length, he will by this means obtain a possible conception of the road on which we plodded all day.

Thanks to a good bed, and a sound night's sleep, I rose on the ensuing morning with no diminution of spirits, and with my physical condition quite unimpaired. A little stiffness in the joints of the hips and knees was all the trace which remained of my yesterday's fatigue; and even that wore away with the first hour's exercise.

At the outlet of the long chain of lakes which stretches far into the heart of this region, we were obliged to wait a few hours for the arrival of the boat which we had engaged, and which was absent on the upper lakes. The time of our delay was profitably employed in taking a fine string of speckled trout from the stream, which here debouches from the lower extremity of the lake. There are few sensations in nature more satisfactory than the gentle titillation of the wrist and elbow, ensuing from the bite of a fine trout; and when the struggle is over and you have him safe in your basket, though you are not indued with the poetic temperament, and may not have an atom of sentiment in your organization, you can hardly suppress a sensation of regret at having destroyed a creature of such rare beauty.

So at least, I think, as I fill my basket; but Newhouse, I am sorry to say, does not share in my weakness. His alimentive instincts are stronger than his idealic; and while I am half disposed to sentimentalize over our prey, he extricates a frying-pan from our luggage, and soon tempts my olfactories with a savory odor, of which, sooth to say, with my appetite sharpened by exercise and abstinence, I am in no wise unappreciative.

By the time we had finished our repast, our boat had arrived; and after securely packing our luggage in the bow and stern, and under the seats, we pushed off from shore, and directed our course toward the upper lakes.

The lakes of the North Woods are a peculiar feature of the region. A chain of small and picturesque sheets of water, eight in number, and connected with each other by shallow channels, extends far into the interior of the wilderness. These lakes are invaluable in the facilities they afford to hunters and trappers and others, wishing to penetrate the heart of the tract; as the transportation of one's self and baggage is rendered comparatively easy, by means of boats. Adventurers in this region can procure a boat at the outlet of the lower lake, and journey upward at their leisure till they find a suitable place for a camp. Those who have walked from the outside world to the landing, will appreciate the value of this arrangement, especially if they have transported their baggage thither on their own shoulders.

The day was drawing to a close when we turned the bow of our boat to the shore, and landed near the foot of lake No. 4 of the series. In our search after a proper location for our camp, we were so fortunate as to find an unoccupied "shanty" of the first quality, of which we lost no time in taking possession.

A "shanty" proper is an institution peculiar to the woods. The most common variety, which the woodsmen erect for temporary use, is made of spruce bark, carefully peeled, so as to preserve the full width, and opened flat like a mammoth shingle. A low frame-work of poles is then constructed, and this bark is so disposed thereupon as to form a dwelling which is nearly impervious to rain. One side of the edifice, however, is always left open, and in front of this the fire is built, which serves to warm the occupants in cool weather. The more aspiring style of shanty, to which ours belonged, is built of logs, halved together at the ends, like a log house; the interstices filled with clay or moss, and the roof covered with bark or split logs. These are intended for more permanent use, and are built by those who regularly frequent certain localities in the woods.

Night drew on. We had barely time to settle ourselves in our new habitation, build our fire, and eat our supper, before darkness overtook us, and we prepared for bed. Our couch was of the most primitive character. A pile of green hemlock boughs, laid upon the bare earth, constituted both bed and bedstead, sheets and coverlets. The only addition to the rather meagre simplicity of this arrangement was a light woolen blanket, for use in an unusually cool night. Newhouse, indeed, had provided himself with an enormous bag, a sort of *cul de sac* of Canton flannel, into which he crept at night very much as a woodchuck ensconces himself in his hole. But I disdained all such artificial appliances. Having turned woodsman, I resolved to make a clean thing of it; and throwing myself upon my rude couch, with a bag of Indian meal for a pillow, in five minutes I was sleeping as sound as though reposing on the downiest of beds, and with the softest of pillows.

My repose, however, was not destined to be uninterrupted. At midnight the chilliness of the air awoke me. I drew my blanket closer around me and tried to compose myself to sleep, but in vain. The novelty of my situation and the unusual sounds which attracted my attention were not at all favorable to slumber. I could hear the distant howling of wolves on the sides of the hill, at the foot of which we were encamped. Then, as I listened, I heard the underbrush crackle, and heavy footsteps tramped though the thicket but a few feet from my head, in the rear of the hut. What was it? A bear? or panther? or wolf? All these animals abound in the North Woods, and the tread was too heavy to have been made by a beast of less magnitude. I reached for my rifle, which stood at my head, and peered steadfastly out into the darkness, but could distinguish nothing. Meanwhile the footsteps had died away in the distance, and my nocturnal visitant had retreated, without deigning to reveal himself. Having by this time become pretty thoroughly awakened, I sprang up, raked together the decaying embers of our last night's fire, piled on a quantity of brush and logs, which created a genial blaze, warming every corner of our rude habitation; then, enveloping myself in my blanket, I slept soundly till awakened by the first beams of the morning sun. Such was my first night in the woods.

The days passed pleasantly in this sylvan retreat. When we were tired of our locality, it was a comparatively easy operation to effect a "change of base." A half hour at any time sufficed to transfer our effects from our habitation to our boat, and another half hour was amply sufficient to establish our *cuisine* and lodging in any locality to which we chose to migrate. Space and time would fail me were I to attempt to describe in detail our multifarious adventures in search of game and trout; how I rowed up and down the lakes trolling for salmon-trout, till the four broad blisters on my right palm, and the three ditto on my left, rendered ample testimony to my proficiency as an oarsman; and how at last, at the close of one pleasant day, we found ourselves securely encamped on a rocky peninsula extending for a mile or two out into the clear waters of Moose Lake.

Moose Lake is an isolated but beautiful sheet of water, lying a mile or two aside from the chain of lakes on which we had hitherto been located. This lake is famed for the abundance and superior quality of its trout; and I was not slow in testing the validity of its reputation in this respect, by catching a fine mess of speckled trout for breakfast on the morning after our arrival, before Newhouse had emerged from his bag of Canton flannel. But as salmon and not speckled trout were the principal objects of our labors, we prepared at once for taking them scientifically. And lest there should be some among my readers who do not clearly apprehend the distinction between the two, I will devote a paragraph or two to their enlightenment.

So few are unacquainted with the common brook or speckled trout, that any description of this superb fish will perhaps be superfluous. The salmon differs from the speckled trout in being more slender in form, and lighter-colored; his flesh rarely assuming so deep and rich a hue as that of the latter, and his spots being more dull. The still, deep water of these wild lakes is his favorite habitat, and there he often grows to the weight of forty or fifty pounds, while the speckled trout is rarely found in water of any considerable depth. The bite of the salmon, too, is materially different from that of the speckled trout. The latter announces his presence by a sharp, eager nibble; while the salmon bites with a sullen, dogged jerk, very much like that of a perch, cat-fish, or Oswego bass. His bite, however, is very sure, and a practiced fisherman will seldom lose the fish that once takes his bait.

The most effective method of capturing the 'salmon-trout with a hook, is, to station a number of buoys in eligible localities, and, previous to fishing, bait them liberally with small fish chopped into pieces as large as the end of one's finger. The salmon, having obtained a taste of the bait, will haunt the place for days afterward; and by baiting the buoys two or three times a day, the fisherman will often obtain six or eight fish from a buoy at a single visit, weighing from one to five pounds each. If small fish for bait are scarce, as is often the case, the buoys can be baited with the inwards of the trout themselves, cut into small pieces with a hatchet on the bottom of the boat.

One pound is about the minimum size of the salmon-trout as they are taken in the northern lakes; and very few smaller are caught. When it is taken into consideration that a single person can manage ten or twelve buoys with a good degree of success, it will be seen that this method of fishing can easily be made profitable as well as pleasurable, to those who are disposed to turn it to account in that way.

Great care is requisite in landing the salmon-trout, or he will break loose from the hook between the water and the boat. From the moment the fish is hooked the line should be kept tight, or he will disengage himself. Pull in your line as rapidly as possible, and your prize will run directly to the surface; and then by taking a dexterous advantage of his momentum, and keeping him carefully clear of the side of the boat, you can throw him clean over on to the bottom with very little outlay of physical force. When captured, insert the sharp point of a knife into the back at the spot where the head joins the body, and he will neither disturb your temper nor entangle your lines by unnecessary floundering. Some amateurs make use of a landing-net; but the practiced sportsman will pronounce this a superfluity.

The recollection of the time passed amid the still loneliness of this beautiful lake will long remain a bright spot in my memory. The passing glimpse of a deer on the distant brink, sipping the clear water in safety, far out of rifle-shot; the occasional shooting of a gull or loon, whose unearthly cry at dusk is forcibly suggestive of a monster not less formidable than a bear or panther; the daily exercise of trout-catching; unlimited rations of trout in every possible shape — trout salmon and trout speckled, trout large and trout small, trout boiled, trout roasted, and trout fried: such is a brief epitome of my life at Moose Lake.

But this wild existence, however pleasurable, must be transitory. Duty recalled me to the world, with a voice too imperative to be disobeyed; and accordingly, having resolved to commence our return journey on the morrow, my companion and I began to pack our baggage in readiness for an early start. How to reduce our effects to light marching order was something of a problem; and while Newhouse was trying to solve the vexed question, I volunteered to "wash the dishes." Our table-service was as follows: One camp-kettle, capacity four quarts, serving the purposes of hot-water boiler, stew-pot, oven, &c.; two tin table-plates; two tin pint basins; two pairs knives and forks; and two iron table-spoons, besides our indispensability, the frying-pan. These I deposited on the shore of the lake, and, making an extempore dish-cloth from a rag which I found among our luggage, I commenced my work. I began with the frying-pan, as being entitled to the most labor, and scrubbed vigorously for what I considered a suitable length of time, but, for some reason, failed to make very sensible progress. The grease adhered pertinaciously;

and the harder I rubbed, the worse it looked. I then bethought myself of commencing with something which would afford an easier task; so I threw aside the frying-pan, and took up one of the tin plates. But here again I experienced a similar difficulty. Rub as hard as I would, the grease obstinately refused to yield to my efforts. By this time, I had begun to think there was something wrong in my way of going to work; so I ceased manipulation, and fell to speculating on the probable cause of my defeat. I had not studied the matter a great while, when it occurred to me that the attempt to wash a greasy dish without either soap or hot water was not an altogether sagacious method of procedure. Having remedied this fundamental error, I experienced no further difficulty, and even congratulated myself on making the discovery unaided. I omitted, however, to mention the circumstance to my companion, partly because my feelings on this point were tender, but mainly because I wished to avoid tempting him into the vice of ridicule — a weakness in which he is at times prone to indulge. He subsequently heard, however, the story of my dish-washing, and, to this day, cannot resist the temptation to start a laugh over it at my expense.

By daylight on the following morning, we were *en route* for home. We had selected and packed for preservation about forty pounds of our choicest fish, and left behind us everything not needed on our return journey. A few hours of rowing brought us to the landing, where we bade farewell to our boat, which had stood us in such good stead. We were now dependent solely on our legs for the transportation of ourselves and effects back to civilization, and we braced ourselves manfully for the task.

As it fell to my lot to carry the said forty pounds of trout, I heroically shouldered my burden, and started in a homeward direction. It was now two o'clock, P. M.; and before we could reach a resting-place, we must traverse those eleven miles of forest which proved so interminably long on our way hither. Certain ominous doubts as to my ability to accomplish the task were carefully thrust aside as irrelevant and not to be entertained.

The sensation, to one who has never before had a load on his shoulders, of a pack of forty pounds' weight placed thereon, is any thing but comfortable; and still less so was the prospect of carrying such a burden over the long and difficult path which lay before us. But circumstances were inexorable: the cross must be borne, and bear it I did, as the sequel will show. By dint of occasionally shifting my load from one point to another on my back, I traversed the first two or three miles quite comfortably. I even began to be jubilant over my supposed capacity as a beast of burden. How great, thought I, will be the shame and confusion of W—— and T—— and H—— (who had striven to cast discredit on my backwoodsmanship), when I relate to them, in full conclave, my triumphant exodus from the wilderness! What, after all, was there in the crossing of the Alps by Napoleon or Hannibal; the passage of the Splugen by Macdonald, or the Rocky Mountains by Frémont; the scaling of the Heights of Abraham by Wolfe; the landing of the Pilgrim Fathers; or any of those achievements about which history makes such an ado — what is there in all these that evinces a greater supremacy of mind over matter, than this march of mine from solitude to civilization with forty pounds of salmon-trout on my back? The greatest deeds are not those which Fame trumpets to posterity. "Full many a flower is born to blush unseen," &c.

But alas for poor, fallible human nature! The spirit indeed was willing, but the flesh seemed likely to prove a failure. At the close of the fifth mile, I felt desperately tired and uncomfortable. Sombre thoughts began to creep over me. What if, after all, my enterprise should not prove a triumph? What if it should result in an ignominious defeat? What if darkness should overtake me, and I should be left exhausted in the forest, a prey to wild beasts? What if the next traveller should find my bones by the way-side, picked clean by remorseless wolves? And, as if to give force to the suggestion, Newhouse, who was a short distance in the rear, shouted, "A wolf! a wolf!" My sporting instincts at once prevailed over my fatigue; and, cocking my rifle, I rushed into the bushes in the direction indicated, just in time to hear the retreating footsteps of the animal dying away among the underbrush. Pursuit was hopeless; but the excitement of the affair revived my drooping energy, and for a short time I trod the lonesome path more lightly.

But this factitious strength was only temporary, and I was soon more tired than ever. So utterly demoralized did I become, that the sight of a noble buck standing directly in my path, but a few paces distant, and gazing at me with his large, lustrous, startled eyes, brimful of wonder, failed to arouse me in the least; and I allowed him to walk leisurely away, unmolested. The only desire of which I was conscious was, an irrepressible longing for shelter and repose, neither of which were near at hand.

It was now nearly dark, and we had yet several miles to travel. Newhouse had loitered a mile or two behind, and I was quite alone. I had long desired to be in the wilderness at night, far from any human being, for the purpose of testing my strength of nerve. I had been curious to know what would be the effect upon me of such a situation, and whether my ordinary equanimity would be in any way disturbed by it. Here was an admirable opportunity to have this question definitively settled; but, unfortunately, I was too tired to indulge in self-examination or thought of any kind, and so allowed the occasion to pass unimproved. About this time, a heavy thunder-cloud, which for some time had been sending forth ominous mutterings, began to discharge its damp contents upon my devoted head. But I was so insufferably weary as to be entirely oblivious of rain, or thought of personal danger. The not unfrequent intimation of the close proximity of of some wild beast caused me no uneasiness, and I could have faced all the animals in the North Woods *en masse* with the most perfect imperturbability. I thought of heaven as a place of rest, and wished I was safely there. I thought of the rude log-hut I had left that morning, and my bed of hemlock boughs, with sensations similar to those with which Adam must have contemplated his lost Paradise. The forty pounds of salmon trout on my shoulders weighed down more heavily than the rocks with which Dante has loaded some of his unfortunate sinners in purgatory. And so I fared slowly on, stopping once in thirty or forty rods to rest, half inclined to throw away my gun and burden, and yet impelled to their preservation by a sort of native tenacity which was unwilling to relax any part of my programme.

By this time it was so dark that eyes were a superfluity. The only method by which I could keep my path was, to be sure that I was safe in the mud. If, at any time, I chanced to set foot on dry land, I concluded at once that I had lost the road; and my progress could only be resumed by groping about until my path was once more wet and miry. I sometimes wonder how I ever succeeded in working my way through such a labyrinth of difficulties; and have gratefully attributed it to the assistance of my guardian angel, who must have been sensible that I was very much in need of his aid.

As all things terrestrial, however, whether good or evil, must at length come to an end, so in due time came the termination of my toilsome journey. After a period which then seemed a great many hours in length, but which a cool retrospect convinces me must have been a much shorter period of time, I staggered out of the woods, and sank exhausted on the bank of Moose River. It is my firm conviction that, if my life had been at stake, I could not have walked another mile. So I lifted up my voice, and shouted vigorously for the ferryman on the opposite side of the river, but I received no reply. It was late at night, and he was doubtless sound asleep long ere this. Then I essayed to awaken him by discharging my rifle; but, on snapping the locks of each barrel, I found that the rain, which was still falling liberally, had so dampened the priming as to make the attempt abortive. I crawled under the shelter of a large hemlock, and finally, after much awkward fumbling in the darkness, succeeded in re-capping my rifle, and in pricking some dry powder into the tubes. This being accomplished, I succeeded in discharging both barrels, and was gratified to hear an answering shout in return. In due time a boat appeared, and I was speedily transferred to comparatively comfortable quarters, where rest and shelter awaited me.

A more forlorn object than myself, as I emerged from the woods on that memorable occasion, it would be difficult to conceive of. Mud, rain, underbrush, and mosquitoes had so thoroughly disguised me that it would have required an intimate acquaintance to have recognized a familiar feature; and, as my powers of description are limited, I will leave the conception of my personal appearance to the imagination of the reader. Thanks, however, to an elastic constitution and a sound night's sleep, the ensuing morning found me comparatively fresh, and in good condition for the completion of my journey.

Thus ended my campaign in the woods. We were now within reach of the appliances of civilization, so that it was an easy matter to reach the nearest railroad station, and avail ourselves of steam locomotion. Despite the numerous vexations and petty hardships which necessarily attended my novitiate in backwoodsmanship, it was, on the whole, a pleasant episode in my life, and one upon which I look back with none but pleasurable emotions. And, although I should not choose to establish my permanent residence in the woods, I can readily sympathize with the attachment to the forest which characterizes those hardy spirits who are " to the manor born."

My narrative has but little to do with trapping or even hunting, and may seem irrelevant in this book. But it relates at least to life in the woods; and, to give any thing like a full view of that subject, it is important to show how such a life seems to the inexperienced. This book may induce many untried youth to enlist in the trapping service; and, in mercy to them, I would give them beforehand some adequate conception of the realities before them, that they may not rush upon the mosquitoes and gnats and wolves and whisky, and long, weary, back-laden journeys, wholly unprepared.

And, after all the tribulations that I have recorded, I came out of the woods with such an appreciation of the discipline of forest life, that I cannot conclude without expressing the hope that the time will come when schools, that prize the health and hardihood that come by gymnastics and military drill, will turn their attention to hunting and trapping as means of education; and, instead of sending occasional detachments of schoolboys in summer on mere aimless pedestrian journeys, or into mock encampments, will annually organize regiments of stalwart youth for penetrating, even in winter, the savage regions far beyond the frontiers of civilization, and doing real service as hunters and trappers of wild beasts — a service, in my opinion, as dignified and disciplinary, to say the least, as war.

TRAVELLING IN A CIRCLE.

By J. P. H.

The little pocket-compass is by no means a contemptible article in the estimation of a woodsman; it has a place in his affections equal, perhaps, to that of his dog or gun, and not only guides him unerringly through the dense and trackless forests, but oftentimes serves him in the capacity of a time-piece. He places it on the muzzle of his gun, and, if after it has become settled, the south end of the needle points directly under the sun, he concludes that it is noon; or, at least, near enough for his purposes. His compass is the most trustworthy servant he has, and it never fails him if he heeds its admonitions. But the inexperienced woodsman is sometimes quite apt to get into a quarrel with his compass, especially when he loses his bearings and gets his head a little turned. Thus, when most needing its aid, he frequently heaps curses upon it, and declares it is "all out of fix." Or he imagines he has come into close proximity to a bed of iron ore, lodestone, or some other wonderful thing that has bewitched his compass. It does not at first occur to him that there can be any thing wrong in his calculations, but he reasons something after the style of the old Indian, who, when he was unable to find his wigwam, vehemently declared, "Indian no lost! Indian here! Wigwam lost!"

It is a noteworthy fact that persons lost in the woods travel in a circuitous course so accurately that they sometimes revolve round to the same point several times within a short period.

T—— told me that he was once travelling in the woods when there was snow on the ground, and all of a sudden came across the track of a man. The track seemed to coincide with his course, so he "struck" into it and followed on, thinking that ere long he might overtake the lonely forester; but, after he had travelled half an hour or more, he, to his surprise, discovered where another track had joined the first. The two travellers, seeming to be of the same mind as himself, were travelling on before him. "Well," said T——, after a brief philosophical parley with himself, "the more the merrier." So he betook himself to his legs and got on somewhat faster than at first. But presently he stood aghast at a third track which "struck" into the trail. Fortunately our hero's native sagacity came to his aid, and he was led to examine the tracks of his neighbors a little more minutely than he had done, and his investigations resulted in the discovery that they were all very much like his own; indeed, so much so that he deemed it perfectly safe and altogether expedient for him to take the back track of the last traveller, and, if possible, make his way out of that scene. He accordingly did so, and in due time found himself "all sound, sir!" as he says, safely landed at the point from which he started when he first entered the forest. He is a little reticent respecting the fate of his three fellow-bushmen; but rumor says they were never heard of more.

I do not remember to have ever heard a satisfactory reason why a lost person travels round in a circle. It has been said by somebody, that people generally step a little farther with the right foot than they do with the left; so, when they have nothing to guide them, the tendency is to bear to the left: thus, in time, they make a circle. But this explanation is not generally accepted. I am satisfied from experience that foresters, when lost, do not always turn to the left, and also that the size of the circle thus made depends very much upon the character of the forest. If it is open, and free from underbrush, one does not deviate from a direct course so much as he would if it were more dense. When a lad, I was connected with an adventure which bears somewhat on the point in question.

Near the head waters of the Chenango River, in New York,

is situated a large swamp, called by the inhabitants of that region "The Great Cedar Swamp." It is eight or ten miles long, and perhaps two or three wide. So boggy is the ground and so impenetrable is the forest for man or beast, I doubt if some parts of it have ever been explored. Neverthless it may be looked upon as a godsend to the surrounding inhabitants; for they are almost exclusively hop-growers, and from this swamp they get an abundant supply of cedar poles, which are gathered in winter, when the ground is more or less frozen.

The Chenango River flows through the whole length of this swamp; yet it is so deep and so sluggish, that the motion of the water is scarcely perceptible, and so crooked, too, that my boyish fancy used to picture a bird trying to fly across it and invariably lighting on the same side from which it started.

At the lower end of this swamp the river leaves the forest, and, losing its mysterious air, breaks into a merry babble, as it hurries away over the stones towards the Susquehanna. Here an old fisherman used occasionally to leave his boat after one of his fishing excursions up the river, and it was the delight of the adventurous youths of the neighboring town to obtain this boat, and penetrate as far as possible into the dark recesses of that solitary swamp.

It was one sunny Sunday when such an opportunity presented itself to me. Two fellows older than myself (one a young man) proposed that we should make an excursion up the river. This was readily agreed to, and we at once possessed ourselves of the boat. We were told, however, that the owner intended to use his boat, and very likely would be after us before we returned. Heedless of all warnings of this kind, we pushed off, and were soon lost among the alders along the stream.

We paddled slowly on for half an hour, with nothing to interrupt our tranquillity but the occasional splash of the muskrat, as he disappeared beneath the black water; or as the spotted turtle, startled at our approach, rolled from off his sunning-place and also disappeared.

We had almost concluded that we were destined to enjoy an undisturbed possession of the boat, when we heard, far away down the river, the stentorian cries of the old fisherman. We at once comprehended the situation of things and knew that he was on our track. We were afraid to go back, and dared not go forward. So we awaited our destiny; nor did we wait long ere we saw his grizzly visage peering through the jungle, demanding our surrender, which demand we immediately and unconditionally complied with. He had a companion with him, and they were both pretty well spent, owing to their hard tramp. We expected to get a ducking, or something worse; but on reaching the shore he only gave us, as I thought, a moderate cursing for not bringing the boat back when he first called to us, for he was sure we heard him. I felt some desire to confess my fault and ask his pardon; but not so, I fear, with my comrades. The big individual was deplorably deficient in his moral department, and unhesitatingly made oath to any thing that his short-sighted nature told him would soonest help him out of difficulty. After laboring much to impress us with the hardships he had encountered in getting up the river, the old fisherman ended by informing us that we might get back the same way he came, or by any other we pleased; he should not ferry us back. Then, to soften the matter a little, he said that a half mile's travel directly away from the river would take us out to the settlement, and thus avoid the difficulty of following down the river.

Now, there is an inexorable law pertaining to human nature, that "every man shall be rewarded according to his works." And I think this old fisherman did not deal with us according to our deserts; therefore, at this point, Providence took the matter in hand.

We plunged in among the tangled bushes as we were directed, congratulating ourselves with our good luck in coming off so easily. But our self-complacency was premature; for after an exhausting tramp of not less than a mile, I should think, instead of arriving at the settlement as we hoped, we all at once found ourselves standing upon the bank of that same deep and ominously dark river:—

"Deep into that darkness peering,
Long I stood there, wondering, fearing."

I longed to see the face of the old fisherman once more, but he was gone and had not left even a ripple on the smooth water. We thought to follow the river down; but, alas! we had forgotten which way was down. We cast sticks into the water, hoping they would indicate to us its course; but their movement was so slight that we questioned whether it was caused by the wind or current. There was nothing left us now but to try our first experiment over again. This we did with the same results as before. My big companion vented his feelings in a shower of oaths; while I thought that any thing would have been more appropriate, under the circumstances, than cursing; for we were all nearly exhausted, so difficult was it to push our way through the tangled swamp-grass and bushes. But we must get out of our terrible situation in some way; so we sallied forth again. After tramping onward for some time, I remember, we came to a partially decayed fish-basket, that had probably been long since left there by some fisherman. Glad to see any thing that reminded us of civilization, we halted to rest a little, and to hold a brief consultation. Here we resolved that, if it was our lot to strike the river again, we would not leave it, but wait for the boat to come down and take us in. With this resolution we again set forth; but our senses had become so bewildered, that, I dare say, we had not travelled ten minutes before we came plump upon that old fish-basket. My big companion was again violent in his expressions; but we scrambled on, not knowing whither we went, until, as chance would have it, we again stood on the bank of the river.

We now climbed a tree, and set our lungs as vigorously at work as our legs had been, calling loudly for the old fisherman. It was not until we had screamed ourselves hoarse that we heard a faint sound far away up the river. It was now his turn to comprehend the condition of things, and after taking his own time for it, he came down to our relief. But, oh, how rejoiced I was to see his craft winding round a bend just above us! He seemed like a father to me: his weather-worn face had a charm about it undiscovered before; no matter if an artful smile did play around his mouth on witnessing our forlorn condition. He spoke kindly to us, and took us aboard his boat, and, after administering some wholesome advice, he soon landed us once more in a civilized region. If I should ever see the old veteran again, I would try to prove to him that I had profited by his advice.

AN EXPEDITION TO THE LAURENTIAN HILLS.

By THEODORE L. PITT.

SEVERAL miles north of the village of Madoc, in Canada West, a traveller, journeying northward, enters upon a section of country to which geologists have given the name of Laurentian Hills. These hills stretch from the Ottawa River to Georgian Bay, and from the neighborhood of Madoc to the region of the Madawaska. This portion of Canada is supposed by geologists to be the oldest land in the world. Here was the primeval continent — the first "dry land" that "appeared" above the all-enveloping ocean, that, in those far-off days of creation, rolled unbroken round the globe. The rocks of this region are the oldest in kind with which man anywhere comes in contact. They are azoic rocks — rocks in which no indications of animal life can be traced. They have no fossils, and if any living creatures existed in the ancient ages in which these rocks were formed, all evidences of their existence have utterly passed away in the geologic revolutions. The country is emphatically a land of hills. They seldom if ever rise to the dignity of mountains, but below this they are of all sizes and shapes. Generally their longer axis is from northeast to southwest. The land appears as if it had once been a vast sea of molten rock lashed into fury by a northwest gale, or the boiling of Plutonic fires, and then in a moment congealed. The region is all underlaid with rock at the depth of a few feet, and it crops out continually. There are visible ledges, vast beds, and bowlders innumerable. Perpendicular cliffs hundreds of feet high are found, sometimes overhanging the clear waters of a lake; at others, the lofty tops of a pine forest. There are great walls of rock piled up, which look as if the Titans of old mythology had worked there in the unknown ages. If one wishes to study rock-work on the largest scale, let him go to the Laurentian Hills and see the backbone of the world. He will see more. He will see the workshop where the continents were made. All the rocks that are now to be seen are but the remnants of what existed in the old ages, hundreds of millions of years ago. They are all ground down and smoothed and rounded by untold cycles of abrasion and disintegration. I can hardly imagine scenery more impressive and suggestive of the mighty power that has worked upon the world in the long, long past.

The Laurentian Hills and valleys are covered with forests of pine, hemlock, hard-wood, cedar, tamarack, &c., and form a paradise for the lumbermen, large companies of whom carry on their operations there. The Canadian government has opened roads running northerly into the forests at intervals of twenty or thirty miles. Settlers have penetrated along these roads and made clearings and erected log-cabins, far into the back country. But it is not a favorable country for farming: the summers are frosty, the winters long and severe, the soil is rocky and shallow. Many deserted cabins are seen, and clearings growing up with forests again. Here and there a section is found where the soil produces fair crops of grain. The greater portion, however, will always remain in woodland, and continue to be one of the best trapping grounds in Canada for years to come. The head waters of several river systems are in this region, and thousands of small streams and lakes abound. The rocks which underlie the country are mostly impervious to water, and the creeks which wind among the hills, wherever they find a basin, fill it and form a lake. These lakes are one of the most interesting characteristics of the country. Their waters are pure and soft. Encircled as they are with woods, the arrangement of the trees around them is a noticeable feature of the landscape. Next to the water is a belt of evergreens, broken rarely in low, marshy places by sections of black ash, or on low, sandy beaches by white birch. Nearest the waters is a fringe of cedars, whose branches droop, and, when the waters are high, touch the waves. Back of the cedars are the hemlocks and pines, and beyond these, on the uplands, the hard-wood timber. In autumn, when the tints are changing, this arrangement forms beautiful pictures. The dark-green of the pines and hemlocks mingles far up the hills, in all picturesque ways, with the splendors of birches, beeches, and maples. The waters of the lake and the cedar fringe form a base to the scene. Over all comes the play of sunshine and shadow.

To this region, in the autumn of 1865, several members of the Oneida Community went on a trapping excursion, under the lead of the old trapper and hunter, Mr. John Hutchins, whose character and adventures have been sketched on previous pages. Their departure from home was announced by the editor of the " Circular " in the following terms: —

" On Monday next, September 25th, an expedition will set out from the Oneida Community for the backwoods of Upper Canada. The object is trapping, and the company go prepared for a six months' campaign in the woods. The expedition consists of —

"John H. Noyes, Perfectionist and Inventor;

" John Hutchins, old Maine trapper and hunter;

" John P. Hutchins, son of the latter, and member of the Oneida Community;

" Theodore L. Pitt, ex-Editor of the ' Circular';

" George Campbell, ex-Financier of the Oneida Community."

The objects of our expedition, more fully stated, were as follows: 1. A practical acquaintance with life in the woods, and its healthful influences; 2. Trapping and acquaintance with trappers; 3. Fur-buying and study of the fur-trade. The programme included within its possibilities a winter campaign in the woods, and an outfit was prepared accordingly. As this outfit was made under the supervision of Messrs. Hutchins and Newhouse, and was the result of their combined wisdom, it is perhaps worth copying, for the benefit of others planning similar expeditions. It was as follows: —

OUTFIT.

Guns; ammunition; fishing tackle; two good salmon

spears; two light axes; two butcher-knives, and one howel or round adze. One hatchet, one pocket-compass, one stout pocket-knife, one double-case watch, a shoulder-basket and a haversack for each man. *Provisions taken from home:* One bushel of beans, two dozen cans of preserved fruits and vegetables, and a few cans of condensed milk. *Clothing, &c.:* One good blanket, one stout suit, two woolen shirts, two pairs of woolen drawers, six pairs of woolen stockings, one pair of camp shoes, one pair of boots, and two pairs of woolen mittens, for each man; scissors, needles, thread, thimbles, wax, patches, &c., in abundance; matches in abundance, in tin safes or bottles, air or water tight; one pocket match-safe for each man. *Cooking utensils:* Two six-quart camp-kettles, two frying-pans, one baking-kettle; tin plates, spoons, knives, forks, basins, coffee-pot and pails. *Miscellaneous:* One drawshave, one hand-saw, one hammer, one inch auger, four gimlets, two lamps and a globe lantern; files, nails, and tacks; pillow-sack and night-cap for each man; sacks for hammock-beds; snow-shoes for each man; fish-oil for bait; ink-stands, pens, and pencils; writing-paper; one dog.

Additional provisions to be taken into the woods were bought at the last village on our route. These consisted of flour, oat-meal, sugar, butter, salt, pepper, &c.

The destination of the party, according to programme, was a point on the Hastings Road, near the head waters of the Trent River. On arrival there, we were to reconnoitre, and, if prospects were unfavorable, go on further north. Mr. Hutchins had trapped in that region several seasons before, and considered it a favorable locality for accomplishing our purposes.

We started from Oneida about noon on the 25th of September, and arrived at McKillican's, sixty miles north of Belleville, on the Hastings Road, the third day after, at midnight. It is sufficient to say of the journey, that we had descended in regular order of travel from the railroad to the steamboat, from the steamboat to the stage-coach, and from the stage-coach to the lumber wagon. The next step was pedestrianism: we had enough of that afterwards. I will say, however, that the traveller on the Hastings Road, after reaching Jordan, sixteen miles beyond Madoc, if he consults his personal comfort, will eschew all other modes of conveyance except those with which nature has furnished him, — his own legs, or perhaps horseback-riding. Even the latter is not the safest operation a man can perform. Hastings Road from Jordan to the York River is truly a "hard road to travel."

McKillican's is the clearing and habitation of Benjamin McKillican, a worthy Scottish Highlander, who, with his family, emigrated from Inverness to Canada many years ago. Nine or ten years since, he settled on the Hastings Road, took up government land and began improvements. He is now seventy years of age; a friendly, hospitable, honest man, and a fine representative of the Scottish faith and earnestness in religion. His family, at the time we were there, consisted of himself and wife; two handsome daughters, who in health, refinement, and industrious activity, were noble specimens of backwoods life; and two younger sons. Our acquaintance and sojourns with this family, first and last, are among the pleasant memories of our expedition.

Seven miles west of McK.'s was Mr. Hutchins's old trapping ground. Four years before, he had left it at sixty years of age, and gone to the war. Those years had made as great changes in the backwoods as in the Southern Confederacy. Other trappers had come in and "occupied the land." Settlers were penetrating the wilds on either hand. Fires had swept through vast tracts of forest. Mink, beaver, and fisher had become less numerous. If we would find good trapping grounds we must go on towards the North Pole, or penetrate many miles into the wilderness, east or west from the Hastings Road. The next morning after our arrival at McK.'s, the question of location was fairly before us. We made inquiries, we sent out scouts, we studied the maps of the country. The result was, the selection of Salmon Lake and the adjacent region, seven miles northeast from McK.'s, as our "camping ground." The locality seemed attractive on the map, being full of lakes and streams. It was said to be out of the range of settlements; was unoccupied by trappers. The choice was between this locality and going on forty or fifty miles to the Madawaska region. The latter was far beyond the range of the white trappers, and occupied by Indians who were unfriendly to intruders. We decided for Salmon Lake.

How to get to Salmon Lake was the next question. There were no roads; at least we could hear of none. There was no navigable river. We shouldered our pack-baskets and rifles, and explored. An old winter lumber-road, which was said to run nearly to the point we wished to reach, was first tried. We followed it two miles and a half, most of the way over burnt and fallen timber, and through a swamp half-leg deep in water, the rain in the mean time coming down in a steady drizzle on our heads. At last we came to an old lumber shanty, and camped for the night. As this shanty was a fair specimen of the lumberman's usual habitation, I will briefly describe it. It was about twenty feet square, seven and a half feet high at the sides, and nine and a half feet at the peak of the roof. Each side was built of five great logs, some of which were two feet in diameter. The roof was made of split logs hollowed into troughs, and placed in this position: ᴗᴗᴗᴗ. All the cracks and holes were compactly filled with moss. The chimney was merely a crib of six-inch sticks laid up log-house fashion from the roof, and placed directly over the centre of the building. It was four or five feet square at the base, and served the double purpose of carrying off the smoke and lighting the shanty. The fire-place was an altar of soil and stones surrounded with timbers, raised a foot or more from the floor, directly under the chimney. There were no windows. Around the sides were two tiers of sleeping-bunks. All through the Canada woods, wherever there is good pine timber, these shanties may be found. They are occupied in winter by twenty or thirty lumbermen, and after the timber is all culled, and transported from the vicinity, are abandoned.

We cleared out the rubbish from the shanty, built a fire,

gathered in great armfuls of balsam and hemlock boughs for beds, ate supper, wrapped our blankets about us, and slept our first night in the Canada woods. Already we had begun to feel a fresh vigor pulsating in our veins as we tramped the virgin soil, drank the pure water, and breathed the perfumed atmosphere of the woods. How new and rich the sensation of tramping all day in the rain and swamp-water, through unknown forests, and lying down at night on evergreen boughs to dream of friends far away!

The next morning, Mr. Hutchins, who had been reconnoitering in a different direction, came up with us and reported he had found a better route. As there was no prospect of reaching the lake short of several days' travel, by this route, and as our provisions were nearly exhausted, we cooked a meal of red squirrels, and retreated. A definite plan was now arranged. A mile and a half east from McK.'s was Bass Lake. From Bass Lake to Salmon Lake there was an outlet five miles long. This outlet was reported navigable with canoes, but no one had voyaged through it for several years. P——, who lived on Bass Lake, said the thing was practicable. We concluded to try it. On an island in Bass Lake grew lofty pines suitable for canoes. P—— was an experienced builder of that kind of craft. We would go to Bass Lake, build canoes, transport our baggage to the shore of that lake, and set sail — paddle, rather — down the "Outlet." We worked cheerfully, happily, and hard for a week; built three canoes, got our baggage across from McK.'s, loaded our vessels, and started.

VOYAGE DOWN THE OUTLET.

It was morning; perhaps we should get to Salmon Lake, four or five miles distant, by nightfall. The mouth of the outlet was shallow and narrow, so that we had to deepen it with pick and shovel the day before. No matter; it would grow deeper. One canoe was fifteen feet long, and thirty inches across the gunwale, carrying three hundred pounds of baggage. Three persons occupied and managed it. The other canoes were small; would carry one man each, and considerable freight.

Gradually, very gradually, the water grew deeper, and the big canoe would occasionally float a rod or two, without much lifting or tugging at the paddle. But it would soon strike a log. If the log was seven or eight inches below the surface of the water, the canoe could be pushed over, by using the paddles as poles, without much difficulty. If the log was nearer the surface, other tactics had to be resorted to. How we finally learned

TO NAVIGATE A BOAT IN A SHALLOW STREAM FULL OF STONES AND LOGS,

is thus told by J. H. N.: —

"It sometimes happens that the trapper, in following his line, or in passing from one lake to another, finds himself with his boat in a small stream, with rocks and fallen trees obstructing his way. The Oneida party, in descending from Bass Lake to Salmon Lake, encountered five miles of this kind of navigation. The creek that connects the two lakes was reduced by drouth to a mere rivulet, with only occasional puddles large enough to float the boats; and though somebody had forced a way through, some years before, by sawing and chopping away logs with incredible heroism and perseverance, much of his labor was lost to us, first, because the low state of the water brought out into bold relief the lower strata of logs, which he had easily sailed over; and, secondly, because hundreds of new trees had fallen across the creek since his descent. Moreover, the beaver dams had all been repaired, and we had to work our way over twelve of them. We estimated by rough guess that the logs we cut through or dragged over numbered about twelve hundred, and the rough rocks (far worse than logs) that we polished with our boat-bottoms were about as many more. In the course of nearly three days' work on these five miles of boating, it may be believed that we learned some practical lessons which it will be useful to record for the benefit of future navigators. We tried two ways of getting along, as people generally do in travelling

"Jordan roads;" namely, first, the dainty, conservative way, and afterwards, when stern necessity had lectured us into an accommodating spirit, the "rough-and-ready" way.

"THE CONSERVATIVE WAY.

"October in the Canada wilderness means November in New York, as we found by the snow-squalls we encountered in those three days. Of course the water was far from being warm; and of course the ex-clergyman, editor, and financier shrank a little from wetting their feet! We were willing from the start to wade in water of moderate depth, say up to the ankle, or anywhere below the tops of our boots; and with only this reservation we worked hard and heroically, and, to say the truth, conquered many obstructions and got along tolerably well; that is to say, at the rate perhaps of a quarter of a mile in half a day. Three of us novices had in charge the big boat, with its load of three or four hundred pounds; and our way was, when we came to a log that could be surmounted without chopping, first to run the bow on as far as we could by a vigorous shove of all hands. Then the man at the bow would step out carefully *on the log*, so as not to take water into his boots, and, the bow being thus lightened, the remainder of the crew could shove it further on. The man on the log could not help much, as his footing was not secure, and he had as much as he could do to look out against wetting his feet, and to find a safe way back to his seat in the boat at the proper time. When we had worked along till the log was under the middle of the boat, the bow man would get in, and the 'midship man would get out, *on the log of course;* and finally, when the balancing crisis was past, and the stern came to be the point of friction, the 'midship man would get in, and the man behind get out, *still on the log*. In this way we kept our feet partially dry, that is, dry as they could be with water soaking through the leather, and running in at cracks; but our progress was very slow. Night overtook us before we had accomplished a quarter of what we had undertaken as a mere afternoon's job; and Heaven only knows whether we should have ever reached Salmon Lake if we had not at last concluded to try —

"THE ROUGH-AND-READY WAY.

"John P. had charge of one of the small boats, and at the same time kept within hailing distance of the large boat, so as to assist the three civilians at the worst pinches. He had seen service of this kind in other days, and knew that the best way was to "take the bull by the horns." He laughed at our policy of keeping the water out of our boots by balancing and teetering on the logs, and set us an example of working on firm footing at the bottom of the creek, without regard to the depth of water. He reasoned and exhorted and scolded; and slowly his radicalism began to prevail over our timidity. The ex-clergyman (otherwise called the inventor) first gave in and went to work in John P.'s fashion, without the fear of wet feet before his eyes. The financier soon followed suit, and the ex-editor, slowly, reluctantly, but finally with a faithful willfulness that beat us all, adopted the simple policy of considering cold water a harmless medium to travel and work in, favorable probably to health by causing reaction. Thenceforth we worked at boat-shoving with free hands and firm feet, and a strenuous heartiness that changed toil into sport, and carried us triumphantly through the most tremendous job of uncivil engineering that three civilians ever undertook. The difference between our first policy and our last was, that we began with trying to keep the water out of our boots, and ended with being contented to keep it out of our breeches pockets!

"After our first conversion to the "rough-and-ready" policy, we had still to learn an important subordinate lesson in regard to the best way of economizing vital heat in dealing with the water in our boots. At first we imagined it was best to get rid of the cold and incumbrance of each bootful we took in as soon as possible; and, for this purpose, at every opportunity we would sit down and lift first one foot and then the other to a position about as high as the head, and let the water run out at the top of the boots, taking care of course to keep the pantaloons out of the reach of the torrent; as, otherwise, what left the boots would run down in the cloth tube to the central and posterior regions of the body. But reflection convinced us that this practice of constantly changing the water in our boots was not wise. A bootful that has been worked in for some time becomes partially warm, and soon ceases to be uncomfortable so far as temperature is concerned. In fact it may be conceived of as a kind of stocking, protecting the feet from the colder water outside, and not easily displaced by what flows in at the top. To turn out this warm water, therefore, at every opportunity, and immediately take a charge of cold water in its place, was a great waste of vital heat, which we finally learned to avoid. Thus we came at last to work right along without paying any special attention to our feet, and found in pursuing this policy true economy of force every way, and no ultimate damage to health or comfort."

The party also learned some other things on this voyage, which the same writer reports as follows: —

"BEAVER DAMS.

"Having opportunity for actual inspection of a great number of beaver dams, we got some new ideas about them. Beavers do literally *cut* down trees and *cut* off logs. Their lower front teeth are really *chisels*. We found one that had dropped out, probably, from the jaw of a superannuated beaver. It was a curved tusk, two or three inches long, and, instead of being pointed, was beveled off at the end as accurately as any chisel, and had a true-cutting edge of a quarter of an inch in breadth. We saw many specimens of their work, which, at a little distance, could hardly be distinguished from axe-cuttings. Boys' hatchet-work would not compare with them for smoothness.

"But the idea that beavers build any thing like a common human dam — namely, a regular log structure or stockade, rising with a steep, definite slope against the stream — is a mistake. Their dams are simply huge deposits of sticks and mud, mixed, and laid, apparently without much order, across the stream. We saw none that raised the water more than about a foot; and sometimes the first notice we had of a dam was from running our boat aground in what had appeared to be deep and smooth water. Neither did we find any confirmation of the popular statement that beavers strengthen their dams by a curve or angle up-stream. Some of the dams we saw were straight, and some curved down-stream, but not one curved or cornered up-stream.

"HOW TO 'SHANTY.'

"When night overtook us in the midst of our boat-dragging, the old trapper would say, 'It is time to shanty.' By this he did not mean that it was time for us to go into a shanty, for there was no shanty within miles of us. He simply meant that it was time for us to prepare for the night. The approved method of 'shantying' in this sense, as we learned it from several experiments under Mr. Hutchins's instruction, shall be minutely described; and ought to be carefully studied by all who are liable to be caught out in the woods in cold weather, with no lodging-place but the ground under the stars.

"A party at work or on the march in the woods ought to stop and prepare for night at least an hour before dark; as the work to be done is not trifling, nor can it be done without light.

"The first matter to be attended to is the selection of a suitable place. Any smooth spot under the trees near your line of march might seem to be good enough; especially if you are tired, and shivering with wet feet and wet clothes, and want fire and supper as soon as possible. But, if you choose thus in a hurry, you may repent. You have a big load of substantial wood to prepare for your night's fire, and you must have reference to this in locating your camp. Soft-wood trees, such as hemlock and cedar, are good for nothing; and you must not think of trusting to dead limbs and brushwood. A fire made of these may boil a pot and give you a momentary comfort; but what you want is a huge, solid log-fire that will take care of you for hours together, and allow you to sleep in peace. You must find a spot where there are hard-wood trees, such as maple, beech, iron-wood, or birch, which you can fell right beside your fire-place. Otherwise you will have to conclude your day's work with some of the hardest lugging that you ever tried. This matter of a good supply of hard, green fire-wood is first in importance. Next to this it is desirable to keep within moderate distance of a stream or spring, as you have the food to cook and the dishes to wash for supper and breakfast, and will need a good deal of water. Lastly, for a good place to sleep on, you must have in front of your fire-place a smooth space, nearly level, sloping perhaps a little toward the fire, and if possible a little lower than the fire, so that the blaze will shine fairly over you and cover you as with a blanket.

"Having chosen your spot, one of the party fells a tree as tall as can be found, and ten inches or a foot through; cuts the trunk into logs eight or ten feet long, and works up the top for small wood. In the mean time another man prepares and drives two stout stakes into the ground at the back of the fire-place, about six feet apart, and four feet high, bracing them from behind with other stakes sloping into notches near their tops. Three of the biggest logs are now placed, one upon another, against the stakes, forming a great wooden chimney-back, three or four feet high. For andirons you find, if possible, two large stones; but, if stones are scarce, you cut a ten-inch hemlock, and, taking two short logs from the butt, place them against your back-logs at right angles to them. On these you lay the fourth of your great hard-wood logs; and thus you have the foundation of your night's fire. While some are making these preparations, others ought to be gathering hemlock bark and dry limbs in great quantities to start the fire, and to enliven it from time to time. Also, if necessary, another hard-wood tree should be felled, that you may have one or two extra logs to put on towards morning.

"The kindling of a fire in the woods, especially in a hard rain, requires some science. A good way is to find a dead cedar or other soft-wood tree that leans to the south. The wood and bark on the sunny side of such a tree is sure to be dry. Split off some strips, and reduce them to fine whitlings with your jack-knife, under your coat or other cover; and, with careful manipulation of matches and kindling stuff, you will soon have a roaring fire under and over the great forestick, that will defy the rain. Hemlock or pine bark, taken from dead trees, is excellent fuel for an incipient fire. But it must be laid on carefully in cob-house fashion, with the outside next the fire. After a while, the furious blaze you have started with light material will get possession of the great green logs, and then the fire will take care of itself for hours. Almost literally it shall be to you a 'wall of fire' through the long cold-night.

"Now hang on the kettle for supper. This is easily done by cutting a pole ten or fifteen feet long, sharpening the large end, and thrusting it obliquely into the ground back of your fire-place, so that the small part will rest on the top back-stick, and the end will project over the fire. A twig left at the proper place will prevent the kettle from slipping.

"All that remains, to make ready for sleep, is to prepare your bed. For this, hemlock or cedar boughs will do; but balsam boughs are the best. The handiest way is to cut down a good-sized balsam-tree near your camp, and strip off its top brush either with your jack-knife or hatchet. This bed-material must not be tumbled into the sleeping-place pell-mell; but must be carefully packed, bough by bough, by thrusting the stick-ends into and under the mass, and leaving the brush-ends to shingle over each other, like the feathers of a bird. If you neglect this, you must expect to roll and groan on hard sticks, instead of sleeping quietly on tree feathers. You sleep, of course, in your blanket, with your boots for your pillow, and with your feet to the fire. If 'the stars look kindly down' upon you, no matter how cold the weather is. You can sleep within the magic circle of that Cyclopean fire, though the water freezes hard in your water-pail at a little distance.

"But what if it rains? Then the party must put their blankets into common stock, extemporize a shelter-tent with one or two of them, and sleep as well as they can under the rest, spread bed-fashion. For the frame-work of the tent you can cut five or six fish-poles, and thrust their large ends obliquely into the ground at the head of your bed, so that they slope up over the place where you are to lie, like the rafters of a roof. You fasten the upper ends with strings to a transverse fish-pole; and then you spread the blankets on the rafters, and fasten them by pinning them to the transverse pole and to each other at the middle edges.

"N. B. — Beware of exposing cotton fixings of any kind to the contingencies of a great open fire, with the winds busy and the sparks flying."

The third day of the voyage, about noon, we reached the open waters of Salmon Lake, and never was a sight more welcome to tired travellers.

SALMON LAKE

Is a beautiful sheet of water, six or eight miles long and from one to two miles wide. So far as we explored, we found it surrounded by an unbroken wilderness, excepting two small clearings formerly made by trappers and two deserted shanties. Two miles from where we located, there was a lumber shanty and a company engaged in the lumber business.

HOW WE LIVED AT SALMON LAKE.

This is told in a letter written by one of the party, as follows: —

"At Bob Holland's old Shanty,
"Salmon Lake, C. W., October 21, 1865.

"Dear Friends, — Human society is, after all, but a great *human body.* The head and trunk and vital organs may be represented by the civilized and enlightened portions of mankind, — those portions where intercommunication is the most close and continuous, where the moving forces are generated, and the highest workings of thought and feeling are developed and educated. But this great human body stretches its hands and feet out into the wilderness, where only the Indian, the pioneer, the trapper, and the lumberman are to be found; and where hardihood, and battle with the elements, the forests, and the animals are the required and the prominent facts of life. Here the circulating fluids move slowly, the lines of communication are far between, and the cuticle is thick and tough. The pulsations of the great heart are felt, but they are minute and feeble. The railroad has afar off given place to the stage-route, the stage-route to the lumber-road, the lumber-road to the blazed foot-path of the trapper and pioneer. The school-house is far beyond the horizon. The newspaper, that indispensability of the interior and superior regions of the body, reaches here only by accident and rarely. The sun here rises over the forest-crowned hills of the east, looks all day long on vast tracts of woodland, on clear-blue lakes wood-encircled, on solitary shanties, where solitary men, or perhaps a man and a woman and some children, try to solve their problems of life; looks through forest-branches perhaps on the dingy form of some solitary trapper, who wanders by shaded streams and sleeps by his log-fire; and then it sets beyond the forest-crowned hills of the west. Here is where the hands and feet of humanity are found as it comes to take possession of the earth. Those extremities are worth coming to see, — worth getting acquainted with, — worth appreciating. 'The eye cannot say unto the hand, "I have no need of thee;" nor again the head to the feet, "I have no need of you."' 'We are all members one of another,' and should 'remember those in bonds,' or in the wilderness and extremities of society, 'as bound with them.'

"BEYOND COCK-CROWING AND THE COW-BELLS.

"An Oneida correspondent raises the query whether we have, after all, got beyond hearing the 'crowing of the rooster or the tinkle of the cow-bells.' Our friends need give themselves no anxiety on this point. The rocks and hills of this region (Salmon Lake) are as free from the sound of the church-going and cow-going bells as the valleys and rocks of Robinson Crusoe's island; and the cry of no fowl more domestic in its habits than the loon ever echoed from these shores. Solitary human beings have sojourned here in former years. The old shanty which we temporarily occupy was once occupied by a trapper noted in these regions. This shanty is eight feet by ten, with an average height of five feet. There is an unfinished shanty of more ambitious proportions a few feet in the rear. On the opposite shore is an unoccupied log-hut. At the other end of the lake there is a new lumber shanty, which is now occupied by twenty or thirty lumbermen. The sound of the great trees falling on the distant hill-sides, reminding one of the reports of far-off cannon, and the occasional appearance of one of the shantymen's red canoes passing under the shadows of the cedars on the eastern shores, are the principal evidences that other human beings are near us.

"ELEVEN DAYS ON SALMON LAKE.

"We have now been at Salmon Lake about eleven days. They have been days of active campaigning. We have had to secure means and routes of regular communication with the outside world, bring up our baggage, select ground for our home-shanty, and commence the building of that structure; had to do what we could in the way of securing a supply of fish, and attend to the daily duties of the camp-kitchen and quartermaster's department. I do not know that the details of any of these operations can be given in a way to make them specially interesting to you. Still there are some things that I will note. First, as to the

"QUARTERMASTER'S DEPARTMENT.

"I judge that it has been seldom that five men (three of them six-footers, or thereabouts) have occupied more limited quarters than have we for the last week. The old shanty which we inhabit measures eight feet by ten on the floor, and is five feet high under the middle of its shed roof. In one corner is a stone fire-place, which discharges its smoke through a square hole in the roof. Between the fire-place and the door is a space about two feet and a half by three, sunk a little lower than the average of the shanty floor, in which the cook can stand to prepare the meals, and in which our shortest man, Mr. Campbell, can stand *upright*. The remainder of the floor is covered with balsam-boughs for a common bed. We can just crowd on to this bed (five of us) at night, by stretching ourselves spoon-fashion, with our heads on a log pillow and our feet to the fire. It is rather a difficult matter for one to turn over without a simultaneous movement of the whole corps. By 'moving careful,' however, and with military precision, the thing can be done. To lie out straight on one's back, between the heels and knees, and other protuberances of the sleepers on either side, is an equally difficult operation. Notwithstanding the smallness of our quarters, we are not troubled with the ventilation question. Our door is an old coat, which swings freely in the breeze, and rather assists the draught of the chimney; besides which, there are various crevices in the walls and roof, where the moss and chinking have tumbled out, that give unimpeded entrance to the air, and exit to the surplus smoke. Across the shanty, just in front of the fire and over the foot of the bed, Mr. N. has placed a seat, which we call the 'deacon's seat.' In front of this, we erect a table at meal-time by placing a single leg under one end of a short hemlock slab, and inserting the other end between the logs of the shanty. It is crowding work to get round at evening and morning, or on rainy days, when baking and cooking are going on, and the table is being set. Yet we manage to keep good-natured, and enjoy it. Even such limited quarters are preferable, in the cool nights and days of late October, to the open camp in the woods, and we have been thankful for their temporary use."

By this time we had our home-shanty about half built, and were contemplating a vigorous trapping campaign. We were looking the long Canada winter in the face, and rejoicing in the prospect of a battle with it. John P. had begun to set traps, and in the course of two nights had caught a fine mink and ten muskrats. We had selected a beautiful location on the north shore of the lake for a winter home. Rowing, spearing fish, felling trees, and shanty building had succeeded to the arduous toils of the voyage through the terrible "Outlet." The signs of game were rather scarce in the immediate vicinity of the lake, but our plans were to run lines of traps far back into the northern woods, where mink, marten, and beaver were supposed to exist in abundance. At this juncture it became evident that the health of our captain was not equal to the execution of the campaign he had planned. For most of the time since reaching McK.'s he had been partially disabled. Now, just as we were building our shanty and preparing for effective trapping, and were relying on him for leadership, he was prostrated for nearly two days, and unable to do any thing. A due consideration of his condition, of the fact that we were all novices in trapping except John P., and of the unfavorable indications of the region as to fur, led us to resolve on a retreat and a "change of base." J. H. N. tells the story of his

LAST DAY IN CAMP,

as follows : —

"I was left alone in camp three or four days on account of a sore hand. In the first place I blistered it by chopping and paddling, and finally it became so bad that I could do neither with any comfort. So I stayed at home to be cook and maid of all work. I had remained there two or three days, leading very much such a life as Robinson Crusoe is reported to have done. The other men were off about two miles, and I had the whole shanty to myself, which was not a very great domain. It was generally perfectly still, — not a sound to be heard. The slightest crackle was a startling event. I would jump up and look out to see what was coming, and perhaps it would prove to be a red squirrel, which would peer in through some hole in the shanty, and watch my movements. Several times a great bird flew over which I was unacquainted with. I learned afterwards that it was a raven. They are very much like crows, only larger, and with a voice somewhat different from that of the crow. In order to get along comfortably I had to talk to myself a great deal. On the last day of my stay, J. P. Hutchins left in my charge certain tasks to be performed. For one thing, having caught ten muskrats, he wanted me to put the skins on stretchers. Then John Hutchins the elder, in the dawn of the morning, when you could hardly distinguish one thing from another, shot an animal which proved to be a *skunk*. It was a large one, covered with fat; and they left it in my charge to get the fat off and try it out for domestic purposes. We had been troubled for the want of light, and on killing the skunk it occurred to them that it was a fine opportunity to get some oil for our lamps. I commenced my day's work by washing up the dishes. By 'dishes' I do not mean such as are found at crockery stores. We had just got our tin plates. (Previously we had eaten off cedar shingles, with wooden spoons.) Then I mended my pantaloons, which had sustained a damage one night before, while I was lying near the fire in one of the Canton-flannel bags that Mr. Newhouse recommended. Just as I was going to sleep I felt something biting or stinging my legs, and, on looking, found that I was on fire. With some difficulty we put it out, after a large hole was burned in the bag, and two small ones in my pantaloons. So, as I said, I proceeded to patch these holes. After that I took hold of the business of making a bag of my blanket. I like the idea of a bag to sleep in, but it ought not to be made of cotton. Mr. Pitt hung up his overalls one night before the fire to dry, and when he got up the next morning only a few little pieces and the buttons were left. We found that cotton clothing about a camp-fire is too liable to get burned up. So I took my woolen blanket and sewed it up into a regular sack, which I liked very much. After that I went through the work of putting the muskrat-skins on the stretchers. Then I went and got the fat off the skunk, and tried it out in one of our spiders or sauce-pans, and made a little tin tunnel and put the oil into a bottle. Then I put the sauce-pan into the fire and heated it red-hot, to take out the odor of the skunk. That was my last work. By this time it was pretty well along in the afternoon. I sat down and began to study.

"It was evident from the failing health of John Hutchins, on whom we had relied as the captain of the expedition, but whose advanced age and former hardships in the army and the woods, by flood and field, now told on him, and from the comparative scarcity of game both for food and fur in the district where we were, that the trapping part of the enterprise would not be made to pay. We had had the advantage of a month's "roughing it" in the woods, and had established communication with frontiermen on their own ground; and it appeared clear that our true course now was to get out of the woods and fall back upon the second object of the expedition, namely, the buying of furs. I accordingly advised a retreat of the party towards the settlements on the Hastings Road, and the next day left myself for the 'States.'"

THE RETREAT.

Two days were spent in repacking our baggage, transporting it across Salmon Lake, and down through Gull Lake to the foot of the latter, and then we were ready to return to McKillican's. We had discovered a new route to Salmon Lake, one by which a greater part of the labor and trouble of the Bass Lake passage might have been avoided. Four miles from our shanty, at the foot of Gull Lake, were Canniff's Mills; and from thence a tolerable road connected with the Hastings Road five miles below McKillican's. We had been unable to learn any thing satisfactory about this route till after we had got to the lake. Our provisions and baggage had been brought round to Canniff's by wagon. They were to go back by the same conveyance. Our baggage being all safely stored in Canniff's mill, we packed our shoulder baskets, shouldered our rifles, and started on a seven-mile tramp through the woods to McKillican's. On arriving at the Hastings Road, we at once began to organize for the fur-buying campaign. Mr. Noyes had gone home. Mr. Hutchins and John P. left soon after for the same destination. Messrs. Campbell and Pitt remained to buy furs. They were soon after joined by Mr. Newhouse, and two months were spent very pleasantly tramping over the rough roads and through the snows. Of this kind of travel the writer performed about four hundred miles. We formed an extensive and pleasant acquaintance with all the leading trappers of the region, who are a class of interesting men. We bought nearly a thousand dollars' worth of furs, the profits on which were not quite sufficient to cover the expenses of the whole campaign. We returned to our Oneida home the last week in December, hearty and strong. In its health-producing results the expedition had paid many fold for all it had otherwise cost. In looking back upon it, in view of all its benefits in this respect, the physical and spiritual heroism which it developed, three of our number at least — the inventor, the ex-financier and the ex-editor — will always remember it with thankfulness. I will conclude my history of the expedition with a dissertation by J. H. N. on the

"MIRAGES OF THE SPORTING WORLD.

"The visions of far-off cities, palaces, gardens, fountains, and lakes that beguile the tired and thirsty pilgrims of the desert are probably but tame and rare illusions compared with those that lure hunters, fishermen, and trappers, or the myriads of men and boys all over the world that would be such, on and on, year after year, in the pursuit of boundless successes that are always looming in the distance, but are never reached. For one, I confess that ever since I was ten years old I have been seeking from time to time, in all directions and by many wearisome excursions, for that paradise of sportsmen where one can bag the nicest game in any quantities "as fast as he can load and fire," or where he can catch bass or trout of any desirable size "as fast as he can put in his hook;" but I *have never found it!* The exact spot has been pointed out again and again by very credible informants; but always, when I have reached it, there has been some mistake about it. Either I had come a few days too soon, or a few days too late; or the desired region was a few miles further on, or off to the right or left, or even back of where I started; or somebody had got in before me, and had just disappeared with the load of luck that I expected; or the weather was wrong; or the time of day was wrong; or I had not the right kind of tools and tackle. Thus in one way or another, as a sportsman, I have never got much beyond moderate luck, with hard work and hard fare; and I have come to the conclusion that the sporting world is full of *mirages*, that ought to be exposed and expounded for the benefit of rising generations.

"I do not believe that my indifferent success is owing altogether to individual bad luck or bad management, but that it is an average sample of general experience. I hear the same story from multitudes of amateurs (told of course in their lucid intervals), and even from old Nimrods. John P. Hutchins said that he "never got through a trapping campaign without wondering at himself that he should be such a fool as to leave a good home and a civilized business to plunge himself into a purgatory of unspeakable hardships for small profits and little sport." And even his father, tough as he is in muscle and story-telling, said nearly the same thing.

"The illusions that cover the sporting world come mostly from the inveterate bragging and exaggerations of sportsmen themselves. The old hunter tells all he can, and more than he can truthfully, of his exploits; and says as little as possible of his failures, and the miseries which his successes cost him. Thus the mirage rises, and they who are deceived by it, in their turn, learn to brag of their exploits and conceal their failures; and so the deception passes on from man to man, and from generation to generation.

"I mean to step out of this practice, and tell some things about our Canada expedition that will tend to sober the expectations of novices, and put them on their guard against inflated reports and promises of sport.

"We went to Canada in full expectation of being able to get plenty of venison and fish for our winter supplies. When we came away, all hopes of getting these provisions had vanished, and we had found it necessary to borrow meat of our neighbors, the lumberers, and were about to send to Montreal for a barrel of mess-pork!

"Our illusions vanished one after another in this fashion. We were told that at Bass Lake we could catch fine, large bass in any quantities, either by drop-line or trolling. We fished patiently with drop-lines at various times for hours together, and got one nibble! We trolled the lake up and down with two boats, and caught one bass of perhaps a pound weight!

"We were told that at Salmon Lake, during a week or ten days after the 8th of October, we should find myriads of salmon-trout on their spawning beds every evening, and could spear boat-loads of them and salt them down for winter use. We had prepared two excellent spears and a jack; and we worked hard to gather "fat pine;" and we laid in a store of salt. But we had no success in finding fish, except on one night, and then only in moderate numbers. All we caught were ten trout, averaging perhaps two pounds apiece, and one fine one of over twelve pounds. We had no occasion to salt them, as five of us easily disposed of them otherwise in the course of a week.

"We were told that we could kill all the deer that we should want for the winter. The understanding was that, just before freezing time, we should lay in our stock. I asked how many deer would probably be a fair supply for the party. The answer was, 'About twenty.' Such were our expectations. The reality was this: Our party had the opportunity of seeing at a distance the chase and killing of *two* deer in Bass Lake, by resident hunters. These were all the deer that were taken in Bass Lake or in Salmon Lake within our sight and hearing, or within our knowledge by rumor, during the whole of our twenty days on the hunting grounds. The dogs were baying frequently, and hunters did their best, but no more deer were taken. We had not the slightest chance of killing any in the usual way by running them into the lakes, as our dog was only a puppy that was more likely to lose himself than to find deer. As to the chance of getting venison by the 'still hunt,' that is by shooting deer in the woods, there was little encouragement, as our party only saw one on land during all our journeyings.

"'But how about *bears?* You didn't kill any, of course, but did you see or hear of any'? Well, I will tell you all about bears. We expected to have something to do with them, and provided ourselves with a couple of Newhouse's famous bear-traps; but we did not set them, and of course did not catch any. We saw scratches on a stump, which Mr. Hutchins pronounced to be the work of a bear's claws made for sport, as a cat airs her hooks sometimes by scratching. One night, when we were camping out, Mr. Pitt heard a terrible noise that he thought bad enough to be a bear's growl; but it proved to be the complaint of an owl. And, to conclude, we had a view — in fact, rather too near a view — of a grisly skeleton of a bear, lying by the side of the path leading from our Crusoe shanty to the lake, — a relic left us by some previous hunter and the ravens. That was the nearest we came to seeing a bear.

"'To cut the matter short, What *did* you shoot'? I killed a partridge and a pigeon. Mr. Pitt killed several red squirrels (which, cooked with some dried beef for want of salt, made an excellent stew). John P. killed some squirrels and a partridge. Mr. Hutchins killed a *skunk*. Besides these, we hit several paper marks, and some we did not hit. This is a true account of our hunting and fishing down to the time of our 'change of base' and my departure for the States.

"A tender conscience and compassion for the inexperienced prompts these confessions. Of course the veterans can do better. They have had their say, and will get more credit than we greenhorns any way. All ears are open to them. As a counterpoise to their exciting stories, we feel bound to leave it as our last word to amateur hunters and trappers, that they should not set their hearts on external success and pleasure, but rather on the benefits to be derived from hard discipline. In that case, we can assure them that they will not be disappointed."

APPENDIX.

HISTORY OF THE NEWHOUSE TRAP.

By G. W. NOYES.

Mount with me, friendly reader, the winged horse of imagination for a trip towards the sunset. Away we speed, by the bustling towns and cities of the West, by the gulfward-rolling Mississippi, by the fertile prairies of Iowa and the plains of Nebraska, by the fringe of squatter settlements that bound civilization in that direction, and by the final hunter's cabin that projects, a faint landmark of repose, into the encircling wilderness. On again five hundred miles further. We are now among the buffaloes; and yet another five hundred in a northwesterly direction places us somewhere in the region of the head waters of three, or perhaps four, great river systems, those of the Missouri, the Columbia, the Saskatchewan, and Mackenzie's River; having their several outlets in the Gulf of Mexico, the Pacific Ocean, the North Atlantic, and the Polar Sea: a wild and solitary place. On one side, snow-capped mountains rise in desolate grandeur to a height of 15,000 feet. Dark forests belt the landscape, where streams, issuing from deep gorges in the hills, break to the level of the plains below. Follow this rocky cañon to where its stream and bed widen into a marsh. We are now among the haunts of the beaver, otter, and mink. We deem ourselves the only human visitants of this remote place. But look! a moccasin track in the sand tells us that some one has been here before us. Its course is toward the margin of yonder sluggish pool; and, as we yet

The Community Works at Willow-Place, Oneida, N.Y.

trace the steps with our eye, click! a clash of steel, and the heavy plunge of an animal in the water, struggling between iron jaws at the end of an iron chain, tell at once the story of the Rocky Mountain trapper and his game.

If not tired with this jaunt, allow a year to pass, and then, on the same handy roadster as before, fly with me a similar journey in the opposite direction. We alight at one of the great European capitals; let it be London. It is night; the glitter of gas and glass around us, the whirl of fashion and the roar of trade, with the miles of crowded pavement that stretch away on every side, almost obliterate the conception of such a thing as rural nature, to say nothing of primitive forest solitude. Here in the aristocratic West End, a mansion door opens; a lady, robed and protected *à la mode* (for the night is cool), and attended by powdered footmen, advances, enters a coroneted carriage, and rolls off to opera or court.

Do you see any connection between these two incidents of antipodal real life? None is obvious, certainly; yet, on noting the lady's costume, a tie of association is at once established; for that London dame this moment presses against her delicate cheek the *fur* of the animal whose death-plunge we heard in the mountain stream of the Northwest. Thus, between my lady the Duchess and the Oregonian trapper, between the Saskatchewan and the Strand, there is a chain of relations of which the middle link, both locally and causatively, is the Oneida Community Trap-Shop. If you had examined the trap whose snap was fatal to the mink on our first flight, and whose spoils you saw adorning European loveliness in our second, you would probably have found stamped on its steel spring the words, "S. Newhouse, Oneida Community, N. Y."

The extraordinary growth of trapping as an occupation within the last ten years, stimulated in part by the remunerative price of furs, and in part by the ever-extending arc of frontier settlements at the West, but still more perhaps by the improvement in the manufacture of traps made by the Community under the supervision of its chief in that department, Mr. Sewell Newhouse, will justify us in giving a sketch of the history of the trap business and of its founder.

Mr. Newhouse is a native of Brattleboro, Vt. His paternal grandfather was an English soldier, who, having been taken prisoner by the Americans at the battle of Bunker Hill, afterwards adopted this country as his home. From Brattleboro the parents of Mr. Newhouse removed during his infancy to Colerain, Mass.; and in 1820, when he was fourteen years old, the family emigrated to Oneida County, N. Y. This central part of the State of New York, if not then an actually new country, retained some of the features of a frontier settlement. The Erie Canal, though it was building, was not finished till several years later; and travel was mainly accomplished by means of stage-coaches, which at some seasons plowed their toilsome way through seas of mud. The large kinds of game, as deer, bears and wolves, were not extinct in the great forest basin of Oneida Lake. Fur-bearing animals and salmon abounded in the streams; and a remnant of the Iroquois Indians, several thousand in number, inhabiting reserved lands in this and the neighboring counties, with their bow-and-arrow proclivities, gave a somewhat primitive cast to the population.

With a stout constitution and a taste for field-sports, drawn perhaps from his English forefathers, Mr. Newhouse found his youth not inaptly placed in such a region. While making the usual school attainments in education, and rendering his share of assistance on the family farm, he also became known as a successful woodsman, wise in the ways of all sorts of game, from wild geese to honey-bees, and from bull-pouts to bears. The instinct of a successful hunter or trapper amounts almost to a sixth sense; and this inevitable tracking faculty which enables one man to detect the signs of game and to seize the strategic point for its capture, which to another are quite unintelligible, was strong in young Newhouse. It is unsafe for a pigeon to alight, or for a muskrat to make an audible plunge, within three miles of such a boy. Vulpine cunning may suffice to elude the common range of observation, but it is no match for the awakened sharpness of the practiced woodsman.

The need of a trapper in a new country is not piano-fortes,

or *cartes de visite*, but *traps*. At seventeen, Mr. Newhouse felt this need, and in the absence of other means of obtaining a supply, he set to work to make them. The iron parts of fifty or more were somewhat rudely fashioned in a blacksmith's shop, and for the steel springs the worn-out blades of old axes were made to serve as material. A mechanic of chance acquaintance showed the young artisan how to temper the springs. The traps thus extemporized proved, on the whole, a success; for they would catch, and what they caught they held. After the season's use, they were sold to neighboring Indians for sixty-two cents apiece, and the making of a new supply was entered upon. These in turn were sold and replaced, and thus the manufacture of " Newhouse Traps " was launched.

During the next twenty years Mr. Newhouse worked at trap-making, sometimes alone and sometimes with a partner or with hired help. The extent of his manufacture was from one to two thousand traps per year, which supplied the local demand, and procured for him a reputation for skill in whatever pertained to wood-craft. During this period he also engaged to some extent in rifle-making; and his amateur productions in this line, being noted for their shooting qualities, were considerably sought after. The working season was generally varied by a trapping excursion to the " Brown Tract " or to Oneida Lake, which improved his practical insight into the details of trapology, and also gave the slightly woody flavor to the man that is observable in his taste and ways. At certain seasons he is still subject to a periodical perturbation, tending towards the North Woods, which, though now but seldom indulged, is a sure sign that he has some time been a liege follower of one of the three ancient Rods.

There are allusions made by the neighbors to feats of strength in wrestling, running, &c., formerly performed by Mr. Newhouse. Such is the traditional anecdote of a thorough taming given by him to one or two big Indians who, in a state of drunken pugnacity, forced a quarrel upon him in the street. But, not having verified these stories, and Mr. Newhouse being himself reticent on such subjects, they may better be left to the keeping of that hazy kind of romance which time gathers about the exploits of the Robin Hoods, Davy Crocketts, and other backwoodsmen of history. We may say that, while clearly the possessor of much muscular power and dexterity to be used in an emergency, Mr. Newhouse is a man of gentle disposition, and is regarded by the remaining red men of his vicinity as their true friend.

The characteristics which Mr. Newhouse possesses as a mechanic are a critical eye, sound judgment of material relations, nicety of hand, and a conscientious attention to the *minutiæ* of any mechanism, on which so often its proper working depends. As a trap-maker, his original idea was to make faultless traps, and nothing could swerve him from this point. His solicitude has been that they should catch game, whether they caught custom or not. The reputation which has come to him on this basis, has made it seem desirable to other manufacturers, in several instances, to pirate his name to give currency to their imitations of the "Newhouse Traps." But this quality of particularity, so valuable in the pursuit of excellence, if not combined with other talents does not always lead to great business success; and the Oneida trap-maker would perhaps have scarcely risen above a local celebrity, but for the introduction of him and his business to a new element of energy and enterprise in the Oneida Community.

THE COMMUNITY "NEWHOUSE TRAP."

The Community established itself at Oneida, about two miles from the residence of Mr. Newhouse, in 1848, and the next summer he and his family entered it as members. For several years after this, but little attention was paid to the trap business. A few dozens were made occasionally by Mr. Newhouse in the old way; but it was not until 1855, under a call for traps from Chicago and New York, that practical interest was first directed to this branch of manufacture, with a view to its extension, by Mr. J. H. Noyes. Arrangements were then made for carrying on the business in a shop fifteen feet by twenty-five. The tools consisted of a common forge and bellows, hand-punch, swaging-mould, anvil, hammer, and file. The shop so established employed about three hands. The next year it was removed to a larger room in a building connected with water-power, and the number of hands was increased. Among them were Leonard F. Dunn, George W. Hamilton, and several other young machinists, who, together with Messrs. Noyes and Newhouse, exercised their inventive powers in devising mechanical appliances to take the place of hand labor in fashioning the different parts of the trap. A power-punch was the first machine introduced, then a rolling apparatus for swaging the jaws. Soon it was found that malleable cast-iron could be used as a substitute for wrought-iron, in several parts of the traps. The brunt of the labor expended had always been in the fabrication of the steel spring, and this was still executed with hammer and anvil wholly by hand. Two stalwart men, with a two-hand sledge and a heavy hammer, reduced the steel to its elementary shape by about one hundred and twenty blows, and it was afterward finished by a long series of lighter manipulations. The attempt was made to bring this part of the work within the grasp of machinery. One by one the difficulties in the way were overcome by the ingenuity of our machinists, until at length the whole process of forming the spring, from its condition as a steel bar to that of the bent, bowed, tempered, and elastic article ready for use, is now executed by machinery almost without the blow of a hammer. The addition of chain-making (also executed mostly by machine power) makes the manufacture of traps and their attachments complete.

The statistics of the business thus extended are in part as follows: Eight sizes of traps are made, for the different grades of animals, from the house-rat to the bear; which have to a great extent superseded the use and importation of foreign traps in this country and Canada. The number of these made at the Community works during the last eight years is over three fourths of a million. The number of hands employed directly has been, in the busiest seasons, about sixty, besides twenty-five or thirty who have found employment elsewhere in supplying the iron castings for traps. The amount of American iron and steel used is over 800,000 pounds annually.

We may add that, to complete their arrangements for carrying on this business to the fullest extent of the possible demand for traps, the Community have built recently a new manufacturing establishment on a water-power about a mile from their former works, which will enable them to more than duplicate their production. A view of the new buildings is given at the beginning of this chapter.

With the progress of improvement in their process of manufacture, the cost and price of traps have correspondingly diminished, so that now the western pioneer or farmer's boy can equip himself with traps of far better quality than the weak and clumsy articles in former use, and at much less price. The influence of these little utensils, now so widely used, on the progress of settlement, civilization and comfort, will occur to every observer. The first invaders of the wilderness must have other resources for immediate support than are offered by the cultivation of the soil. These are present in the valuable peltries of fur-bearing animals which are the occupants of the soil in advance of man. Hence the trap for securing them, going before the axe and the plow, forms the prow with which iron-clad civilization is pushing back barbaric solitude; causing the bear and beaver to give place to the wheat-field, the library, and the piano. Wisconsin might, not inappropriately, adopt the steel-trap into her coat-of-arms; and those other rising empires of the West — Kansas, Colorado, Nevada, and golden Idaho — have been in their germ and infancy suckled, not like juvenile Rome by a wolf, but by what future story will call the noted wolf-catcher of their times, — the Oneida Community "Newhouse Trap."

DESCRIPTION OF THE NEWHOUSE TRAP.

THERE are eight different sizes of the Newhouse Trap, adapted to the capture of all kinds of animals, from the house rat to the grizzly bear.

No. 0.

The smallest size having but recently been introduced into the series, is designated as No. 0, and is called the RAT TRAP. It has a single spring, and the jaws spread, when set, three inches and a half. It is designed for the house rat, but is strong enough to hold the muskrat.

No. 1.

No. 1 is called the MUSKRAT TRAP. It has one spring, and the jaws spread four inches. It is adapted to the capture of the mink, marten, and all the smaller fur-bearing animals.

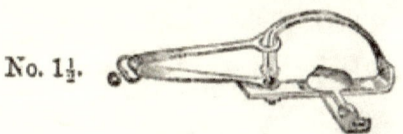
No. 1½.

No. 1½ (also recently introduced), is called the MINK TRAP. It has but one spring; and the jaws spread four inches and seven-eighths. It is strong enough for the fox or fisher.

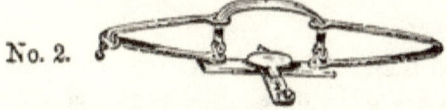
No. 2.

No. 2 is called the FOX TRAP. It has two springs, and the jaws spread four inches and seven-eighths. It is strong enough for the fisher or even the otter. Trappers sometimes have ordered this size with single instead of double springs. No. 1½ is intended to meet such demands.

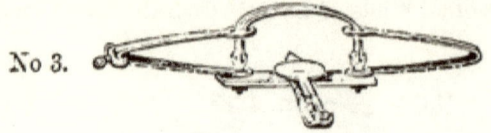

No. 3.

No. 3 is called the OTTER TRAP. The jaws spread five inches and a half. It will hold any of the medium-sized animals, such as the beaver, the badger, the raccoon, the opossum and the wild-cat.

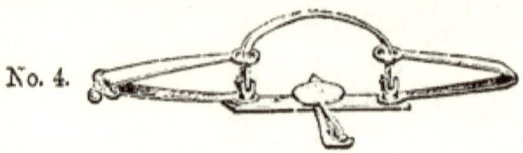

No. 4.

No. 4 is called the BEAVER TRAP. The jaws spread six inches and a half. It is adapted to the wolf or the lynx.— Extra sets of jaws with teeth constructed expressly for taking deer, are made to fit this trap, and can be had separately, or may be inserted in the place of the ordinary jaws.

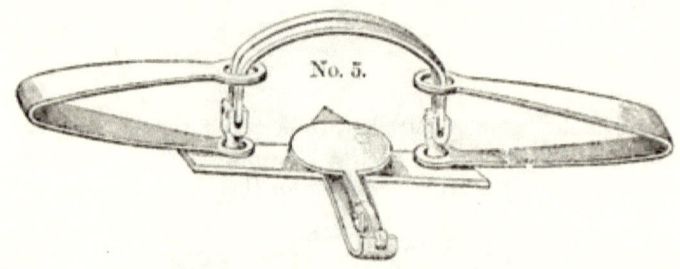

No. 5.

No. 5 is called the SMALL BEAR TRAP. The jaws spread eleven inches and three-fourths. The weight of each spring is two pounds and ten ounces, and the weight of the whole trap is seventeen pounds. It is adapted to the common black bear, the panther, and most of the large animals found this side of the Rocky Mountains.

All these traps are furnished with swivels, and if desired, with chains.

DESCRIPTION OF THE NEWHOUSE TRAP.

THE GREAT BEAR TAMER!

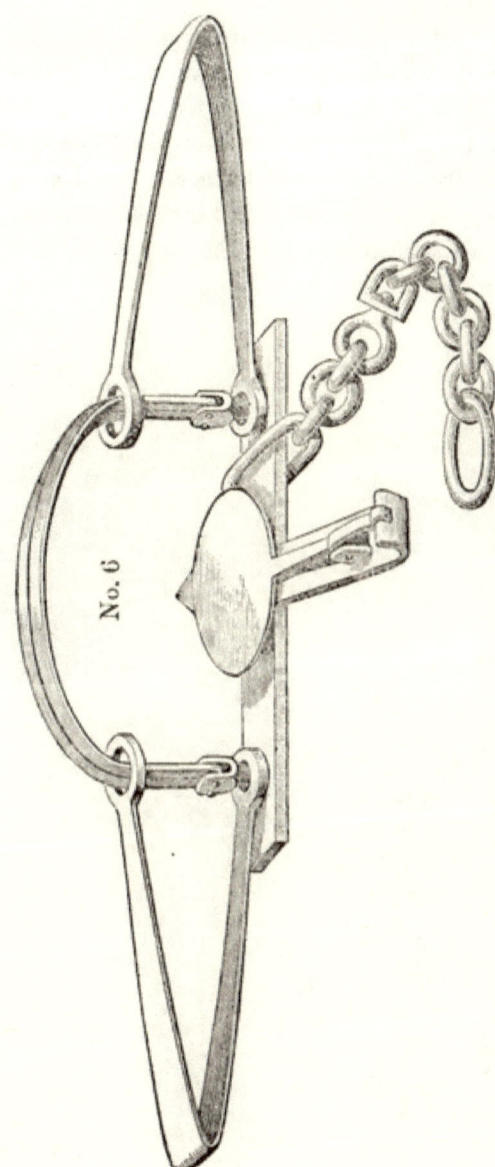

No. 6 is called th GREAT BEAR TRAP The jaws spread sixteen inches. The weight of each spring is six lbs.; weight of the trap, with chain, forty-two lbs. This is the trap for the Moose and the Grizzly Bear. Its tremendous power of taming wild beasts is already known in the mountains of California; but it ought to be known, and we trust it will be, ere long, in all parts of the world. Its use need not be confined to the capture of animals for their furs. In the interest of civilization, it ought to go wherever ferocious animals exclude man from the soil. India, in all her jungles, needs it to exterminate the Tiger. Africa needs it in her long battle with the Lion. South America needs it for grappling with the Jaguar and the Boa Constrictor. There is not an animal living that can defy it, unless it is the Elephant, whose foot may be too large for it; and even the Elephant, taken by the trunk, would have to succumb. It is safer, and far more sure and effectual than fire-arms in encounters with any of these monsters; and ought to be put in the very front of the battle of man against the savages of the forest and the desert.

CONCLUSION.

A LEARNER of any art requires good rules, good examples, good tools to work with, and good practice. Our task has been to furnish the first three of these requisites for mastering the art of Trapping In the first part of this book Mr. Newhouse gives good rules. In the Narratives the learner will find good examples. In the Appendix we show him where he may find good tools. The good practice must be the work of his own genius and resolution.

www.ingramcontent.com/pod-product-compliance
Lightning Source LLC
Chambersburg PA
CBHW031928230426
43672CB00010B/1859